Thin-Layer Chromatography
with Flame Ionization Detection

W0107122

Mojmír Ranný

Research Institute of Fat Industry,
Department of Organic Technology, Institute of Chemical Technology,
Prague, Czechoslovakia

THIN-LAYER CHROMATOGRAPHY WITH FLAME IONIZATION DETECTION

D. REIDEL PUBLISHING COMPANY

A MEMBER OF THE KLUWER  ACADEMIC PUBLISHERS GROUP

DORDRECHT / BOSTON / LANCASTER / TOKYO

Library of Congress Cataloging-in-Publication Data

Ranný, M. (Mojmír), 1925—
  Thin-layer chromatography with flame ionization detection.

  Bibliography: p.
  Includes index.
  1. Thin layer chromatography. I. Title.
  QD79. C8R36 1986   543′. 08956   85-14455

  ISBN-13: 978-94-010-8159-7   e-ISBN-13: 978-94-009-3705-5
  DOI: 10.1007/978-94-009-3705-5

Scientific Editor
Doc. Ing. Jaroslav Janák, DrSc.
Corresponding Member of the Czechoslovak Academy of Sciences

Reviewer
RNDr. PhMr. Ladislav Novotný, DrSc.

Published by D. Reidel Publishing Company,
P. O. Box 17, 3300 AA Dordrecht, Holland,
in co-edition with
Academia, Publishing House of the Czechoslovak Academy of Sciences, Prague.

Sold and distributed in the U.S.A. and Canada by
Kluwer Academic Publishers,
101 Philip Drive, Assinippi Park, Norwell, MA 02061, U.S.A.

Sold and distributed in Albania, Bulgaria, China, Czechoslovakia,
Cuba, German Democratic Republic, Hungary, Mongolia, Northern
Korea, Poland, Rumania, U.S.S.R., Vietnam, and Yugoslavia by
Academia, Publishing House of the Czechoslovak Academy of Sciences,
Prague, Czechoslovakia,

Sold and distributed in all remaining countries by
Kluwer Academic Publishers Group,
P. O. Box 322, 3300 AH Dordrecht, Holland.

Contents

6

Introduction

Thin-layer chromatography (TLC) has become a common and much favoured separation technique in laboratories in widely varied fields in recent years. Much of the credit for the introduction of this technique into analytical practice at the end of the 1950s is due to E. Stahl[1,2]. This method is simple and is characterized by high separation ability and sufficient sensitivity[3]; however, some analysts feel that it has passed the peak in its development and will gradually be replaced by the more modern high-performance liquid chromatography (HPLC). This is undoubtedly a very important analytical technique utilizing the specific separation properties of a large number of sorbents and the possibility of regulating the flow-rate of the mobile phase by adjusting the pressure[4]. Standardization of the experimental conditions is simpler in HPLC than in TLC, where the activity of the sorbent and flow-rate of the eluent in the thin layer depend markedly on the relative humidity of the laboratory atmosphere and on the composition of the gaseous phase in the elution chamber. In addition, systems for quantitative detection of the separated zones are better developed for HPLC than for classical TLC, where, until recently, cumbersome and often even insufficiently reproducible chemical or gravimetric analysis of the extracts of scraped-off spots or densitometry of the separated zones, located first by pyrolysis or reaction with suitable detection agents, were the predominant determination methods[5].

In spite of these advantages, however, HPLC is also not an ideal, universally applicable chromatographic technique.

The separation ability of the columns is usually greater than that of common thin layers; however, the packing is relatively expensive and its separation ability unfortunately decreases quite rapidly as a result of irreversible adsorption of the components of the analysed mixture. The decreased universality and often also lower efficiency of liquid chromatography compared with thin-layer chromatography, is a result of a limited choice of composition of the mobile phase, because of problems encountered in the detection of binary mixtures. The response to the liquid eluent in the detector must be much smaller than that for the test component; otherwise, the selectivity of the detector is insufficient. In this sense, TLC is more flexible, as detection system is not connected directly to the chromatographic system. This flexibility and simplicity of TLC makes it an analytical tool which is just as valuable as HPLC.

Thin-layer chromatography is continually developing in two basic directions. To begin with, it is necessary to increase the separation efficiency of the system by selection of an adsorbent with small grain size, forced mobile phase flow and standardization of conditions during elution and detection. This 'high-performance thin-layer chromatography' (HPTLC) can be used to separate more than 20 component mixture in a single run and, after suitable detection, quantitatively analyse the components spectrophotometrically with quite good reproducibility[6-8].

An interesting modification of HPTLC is OPLC (overpressured layer chromatography)[8]. In this system, the layer of sorbent is covered by a flexible, non-porous, plastic membrane (cushion system) and the mobile phase is forced through the layer similarly to HPLC using a micropump. This promising procedure has a certain disadvantage in the rather high price of commercially prepared layers which, like classical plate layers, can be used in only one single chromatographic process.

A second, technically more advanced step in the development of TLC is chromatography on permanent rod-shaped layers, whose mechanical and chemical properties permit detection of the separated components in an ionization detector. This method, termed TLC-FID, is based on work carried out in the late 1960s in which the principles of gas chromatography were utilized to a certain degree. The sample was separated on a long, narrow, thin-layer plate and this was placed in a quartz tube. The zones were gradually pyrolysed and the gaseous products stripped by a stream of inert gas into a detector[9-12]. Significant contributions to the development of the method were made by Padley[13] and Szakasits[14], who passed the rod or strip-shaped thin layers with the separated components directly through the flame of an FID.

The ideas of Padley were further developed by Okumura and Kadano, who published patents for the Japanese company Shionogi in 1971 to 1974 on a procedure for manufacturing a reusable TLC rod with a sintered silica gel layer, suitable for flame-ionization detection [15,16]. These layers, commercially known as Chromarods, have good separation ability, are resistant to strong acids, permit elution in highly polar solvents including water and, most important, can be reactivated in hydrogen flame.

In the few years since its introduction, TLC-FID has become an important quantitative method (comparable to GLC or HPLC) and is used in practice for separation of chemical substances especially in medicine and biology, lipid chemistry and the characterization of crude oils. Because of its high sample throughput, it is particularly useful for series quantitative separations of substances that cannot be analysed by gas chromatography because of their low volatility or because of the irreversible adsorption of components on the column.

In this book, an attempt has been made to provide basic information on this method and on its application possibilities, on the basis of experience which the author has gained together with his coworkers during more than five years of application to the analysis of natural and synthetic organic substances[17]. The work of other authors published in the scientific literature and by equipment manufactuers has also been incorporated.

For ease of reference, Part 1 of the book (General and Methodical) lists the theoretical relationships applicable to TLC. This is followed by a brief description of instrumental techniques and instructions, which it is hoped will assist the new user of TLC-FID to avoid the unnecessary errors which almost always occur on initial acquaintance with a new analytical procedure.

In Part 2 specialized sections are devoted to the application of TLC-FID in various scientific and industrial fields; the usual approach of classification according to the chemical structure of the analysed substances has been maintained. To assist in finding a suitable separation procedure, tables are included describing the composition of elution systems, together with the corresponding sequence of the individual components on the chromatogram. Wherever possible, detailed procedures are included for the preparation of samples for analysis and information on the reproducibility of the method is given. The lengths of the individual sections vary considerably and, to a certain degree, these differences reflect the success or failure of the application of TLC-FID in the given field.

Chromatography on Chromarod layers can readily be carried out by analytical chemists who have had experience with classical TLC. However, it is not a perfect method that is capable of solving all the problems of chromatographic separation. It should be borne in mind that this is a new, flexible and developing technique, with gradual improvement of the quality of permanent layers and detection systems. In this area, the greatest improvements are expected from the combination of TLC with selective flame-ionization detection (FTID, Flame Thermionic Detector), with a higher sensitivity for nitrogen and chlorine atoms and especially from the application of flame-emission detection (FED) with linear response to carbon in a broad concentration range. Using special filters, selective response can be obtained, for example for sulphur and phosphorus. These new advances are discussed in Part 3.

The development of this method is far from complete and it is thus highly probable that some of the conclusions that were valid when this manuscript was written will be of lesser importance by the time it is published. Nonetheless, the author believes that this book will be a useful handbook for those who have decided to employ this attractive technique in laboratory practice.

List of Symbols

A	alumina (Chromarod A)
AAS	atomic adsorption spectroscopy
A/D	analogue to digital
Ae	surface occupied by an eluent molecule (Eq. 40)
A_i	area of peak i (Eqs. 45–47, 49–53)
ala	β-alanine
A_m	cross-sectional area of the mobile phase (Eqs. 3, 5, 7)
AOCS	American Oil Chemist's Society
as	subscript refers to the sample analysed
arg	arginine
A_s	cross-sectional area of the stationary phase (Eqs. 3, 5, 7)
A_{st}	peak area of the standard (Eqs. 49–53)
ASTM	American Society of Testing Materials
av.	average
b	the width of the Gaussian curve at the base (Eq. 26)
$b_{0.5}$	the width at a half-height of the Gaussian curve (Eq. 26)
B_{SA}	amide of nicotinic acid
B_{SK}	nicotinic acid
BDP	bis(diacylglycerol)phosphate
b_F	the width of the Gaussian curve at the front (Eq. 39)
BHA	butyl hydroxyanisole
BHT	butyl hydroxytoluene
BMP	bis(monoacylglycerol)phosphate
BPMC	2-(methylpropyl)phenylmethylcarbamate
b_s	the width of the Gaussian curve at the start (Eq. 39)
c	solute concentration (Eq. 17, 18)
C	cholesterol
C	vitamin C
CA	cholic acid
CD	chenodeoxycholic acid

CE	cholesterol ester
c_i	concentration of component i (Eqs. 48, 51)
c_m	equilibrium concentration in the mobile phase (Eq. 1, 4, 11)
c_{max}	maximum concentration (Eqs. 17–19)
CMSA	chlormethanesulphonanilide
c_s	equilibrium concentration in the stationary phase (Eqs. 1, 4, 11)
c_{st}	concentration of the standard solution (Eq. 51)
CT	total cholesterol
CV	coefficient of variation
$c_{x,y}$	concentration of the solute at point x, y (Eq. 19)
D_2	vitamin D_2
DC	deoxycholic acid
DDS	diaminodiphenylsulphone
DG	diacyglycerol
D_i	diffussion coefficient (Eqs. 15–17, 20–23, 29–31, 33–36)
di-BHA	dibutyl hydroxyanisole
dp	grain diameter (Eqs. 23, 28, 30, 31, 33–36)
DPG	diphosphatidylglycerol
DT	developing tank
E	vitamins E, tocopherols
EDTA	ethylene diaminotetraacetic acid
EO	ethylene oxide
Eq	equation
Eqs.	equations
ES	ethyl stearate
E_{sa}	adsorption energy (Eq. 40)
Et-CMSA	N(ethoxycarboxymethyl)chloromethanesulphonanilide
F	front of the chromatogram or scanning
FA	fatty acid
FED	flame emission detector (detection)
FID	flame ionization detector (detection)
F_m	association factor (Eq. 15)
FTID	flame thermionic detector (detection)
G	glycerol
GCA	glycocholic acid
GCD	glycochenodeoxycholic acid
GDC	glycodeoxycholic acid
GLC	glycolithocholic acid

GLC	gas-liquid chromatography
GPA	glycerophosphoric acid
GUD	glycoursodeoxycholic acid
h	reduced plate height (Eq. 32)
H	height equivalent to a theoretical plate (Eq. 25)
$\bar{H}$	average plate height (Eqs. 26, 31, 34)
HAFID	hydrogen atmosphere FID
HDF	solvent system n-hexane-diethyl ether-formic acid
HDL	high density lipoproteins
HETP	height equivalent to a theoretical plate
his	histidine
HPLC	high-performance liquid chromatography
HPLP	hyperlipoproteinemia
HPTLC	high-performance thin-layer chromatography
iso-PCA	1-phenyl-4-chloro-5-aminopyridazone(6)
IS	internal standard
k	capacity ratio (Eqs. 2, 3, 5, 6, 20, 22)
K	distribution constant (Eq. 1, 3–5, 7, 11, 38)
k_1	constant (Eq. 14, 43)
k_2	parameter (Eqs. 43, 43a)
K_F	correction factor (Eqs. 46–49, 51, 52)
lan	lanoline
LAS	n-alkylbenzenesulphonate
LC	lithocholic acid
LC	column chromatography
LDL	low density lipoproteins
LPA	lysophosphatidic acid
LPC	lysophosphatidylcholine
LPE	lysophosphatidylethanolamine
m	subscript which refers to the mobile phase
m_a	amount of the sample applied (Eq. 22)
m_{as}	sample mass (Eqs. 52, 55)
MDP	monoacylglycerol-diacylglycerolphosphate
ME	methyl ester of fatty acid
met	methionine
MG	monoacylglycerol
m_i	amount of component i in the sample (Eqs. 49, 50, 52, 54, 55)

M_i	mass of component i in the calibration mixture (Eqs. 53, 54)
m_m	amount of solute in the liquid phase (Eqs. 2, 4)
M_m	molecular mass of the mobile phase (Eq. 15)
m_s	amount of the solute in the stationary phase (Eqs. 2, 4)
m_{st}	amount of the internal standard (Eqs. 49, 50, 52, 54)
M_{st}	mass of the internal standard in the calibration mixture (Eqs. 53, 54)

n	number of plates (Eqs. 24, 25, 38)
n_1	number of plates per unit length (Eqs. 24, 25)
n_b	surface area of the sorbed molecule of the stronger eluent (Eq. 41)
n_{el}	number of elutions (Eq. 44)
NHA	Newman-Howells Associates Ltd., Winchester, England
NL	neutral lipids

| O | origin |
| OPLC | overpressured layer chromatography |

PA	phosphatidic acid
PC	phosphatidylcholine
PCA	1-phenyl-4-amino-5-chloropyridazone (6)
PCC	1-phenyl-4,5-dichloropyridazone
PE	phosphatidylethanolamine
PG	phosphatidylglycerol
PGL	propylene glycol
phe	phenylalanine
P_i	percentage of component i (Eqs. 45–48, 55)
PI	phosphatidylinositol
PL	phospholipids

| PO | paraffin oil |
| PS | phosphatidylserine |

r	radius (Eqs. 14, 43a)
RAM	random access memory
RDS	respiratory distress syndrome
rev	revolutions
R_F	retardation factor (Eqs. 4–9, 20–22, 37, 38, 44)
ROH	fatty alcohol
R_s	resolution (Eqs. 37, 38)
RW_1	4-methylmorpholine hydrochloride
RW_3	4, 4-dimethylmorpholine chloride

| S | soap |

SARA	saturates, aromatics, resins, asphaltines
SE	sterol esters
ser	serine
SM	sphingomyelin
SN	separation number (Eq. 39)
SQ	squalen
st	subscript which refers to the standard
ST	sterols

t	time (Eqs. 12, 16, 17, 19, 20–22)
T	absolute temperature (Eq. 15)
T	tocol
TAK	triacontane
TCA	taurocholic acid
TCE	tetrachlorethane
TCD	taurodeoxycholic acid
TG	triacylglycerol
THF	tetrahydrofuran
TLC	thin-layer chromatography
TLC-FID	thin-layer chromatography with flame ionization detection
try	tryptophane
TUD	tauroursodeoxycholic acid
tyr	tyrosine

u	velocity of mobile phase (Eqs. 29, 30, 31, 33)
UDCA	ursodeoxycholic acid
u_F	front velocity (Eq. 13)
UV	ultraviolet
$\bar{u}_{zF}$	average velocity of the front (Eq. 14)

v	velocity (Eqs. 19, 20, 22)
V_i	molar volume (Eqs. 15, 42)
V_m	volume of the mobile phase (Eqs. 3, 4, 7, 10, 14)
V_s	volume of the stationary phase (Eqs. 3, 4, 7, 14)
V_{st}	volume of the standard solution (Eq. 51)
V_v	volume of the liquid phase (Eqs. 10, 14)

W	wax

X_i	mole fraction of component i (Eqs. 41, 42, 42a)

z_F	migration distance of the front (Eqs. 8, 12–14, 24, 25, 34, 35, 37, 39)
z_o	distance from the origin to the surface of the mobile phase (Eqs. 8, 24, 25, 34, 35, 37, 39)
z_x	the path length (Eqs. 26, 37)
α	activity parameter (Eq. 41)
γ	surface tension (Eqs. 14, 43, 43a)
ε^o	elution strength (Eqs. 40, 41)
η	viscosity of the mobile phase (Eqs. 14, 15, 43, 43a)
Θ	correction factor (Eqs. 8, 9)
$\varkappa$	flow constant (Eqs. 12–14, 34–36, 43, 43a)
λ	factor dependent on the geometry of the layer (Eqs. 28, 31, 34)
v	reduced velocity of the mobile phase (Eqs. 20, 22, 32, 33)
ξ_v	correction factor (Eqs. 9, 10, 43a)
σ	half-width of the Gaussian curve (Eqs. 16, 18–21, 26, 37)
τ	parameter (Eqs. 20, 22, 23)
ψ	factor including the resistance of the layer to free molecular diffusion (Eqs. 29, 31, 34, 35)
ω	factor characterizing the type of sorbent (Eqs. 30, 31, 34, 35)

Part 1 General and Methodical

1.1 Principles of Thin-Layer Chromatography and Basic Quantities in the Chromatographic Process

Theoretical derivation of relationships valid for TLC (and also for paper chromatography) is far more difficult than for other chromatographic procedures; one of the main complications is the change in the flow-rate of the mobile phase with increasing distance of the front from the start, and the consequent decrease in the ratio of the volumes of the mobile and stationary phases with an increase in the length of the chromatographic path. In addition, in contrast to GLC and HPLC, the separation is strongly affected by the composition of the gaseous phase in the elution chamber and relative humidity of the laboratory atmosphere.

The basic theory of chromatographic processes was developed in the 1960s and early 1970s, especially by Giddings[18], Snyder [19] and Novák and Janák[20]. However, the book by Geiss, published in 1972[21], can still be considered as the basic work on the significance of the individual parameters in thin-layer chromatography. The work of Kaiser et al.[6,7], also made an interesting contribution to the theory of TLC.

This chapter will review the theoretical basis of TLC only so far as is necessary for the reader to grasp some of the problems encountered in TLC-FID, to be discussed subsequently in the rest of Part 1. The brevity of this treatment is deliberate, because this is primarily intended as a laboratory handbook.

Like other chromatographic methods, thin-layer chromatography is based on the different distribution of the separated substances (solutes) between two immiscible phases, one of which is moving. The second stationary phase may be either the solid phase of the sorbent — in which case adsorption chromatography is involved — or another liquid phase, immobilized on a solid support; this is then called partition chromatography. Both of these procedures can be involved in TLC. A typical example is chromatography of a sample of mixed liquids, containing solvents with very different affinities for the stationary phase. During the flow through the layer, the mobile phase gradually becomes depleted in the polar component, which is more strongly bound to the sorbent and forms a stationary liquid phase. A secondary front is formed in regions with minimal concentration of this component, below which the analysed substance is distributed between the liquid stationary phase and the flowing system, and above which typical adsorption chromatography occurs between the solid adsorbent and the flowing liquid phase, depleted in the polar component. Displacement chromatography is a special type of adsorption chromatography and is used when the solute is bound to the sorbent by weaker forces than the polar component of the mobile phase; this then replaces the solute in the surface layer and forces it ahead to a position, usually immediately below the subordinate front. This process is characterized by relatively narrow zones on layers of the Chromarod type, or narrow bands on planar thin layers.

The chromatographic process is influenced primarily by the difference in the distribution of the separated components between the stationary and mobile phases, by the difference in their transport rates in the eluent and finally by their diffusion rates in the two phases.

1.1.1 Distribution of the Solute Between the Stationary and Mobile Phases; the Retardation Factor

The distribution of the solute between the two phases is characterized by distribution constant K, defined as the quotient of the equilibrium concentrations of the substance in the stationary and mobile phases

$$K = \frac{c_s}{c_m} \tag{1}$$

Components of the analysed sample can be separated chromatographically only by assuming that their distribution constants are not very different in the given chromatographic system.

A further important quantity is the ratio between the amount of solute in the stationary (m_s) and liquid (m_m) phases, called the capacity ratio:

$$k = \frac{m_s}{m_m} \tag{2}$$

It follows from these two equations that

$$k = K\frac{V_s}{V_m} = K\frac{A_s}{A_m} \tag{3}$$

where V_s and V_m are the volumes of the stationary and mobile phases, respectively, and A_s and A_m are the cross-sectional areas of the particular phases.

An important relationship for TLC (the Martin-Synge Equation)[21,22] can be derived from Eqs. (2) and (3):

$$R_F = \frac{m_m}{m_m + m_s} = \frac{c_m V_m}{c_m V_m + c_s V_s} = \frac{V_m}{V_m + K V_s} \tag{4}$$

where R_F is called the retardation factor (real R_F value) and expresses the fraction of the solute in the mobile phase. Modification of this quantity yields the dependence of R_F on the capacity ratio,

$$R_F = \frac{1}{1 + K(A_s/A_m)} = \frac{1}{1 + k} \tag{5}$$

and thus

$$k = \frac{1 - R_F}{R_F} \tag{6}$$

A substance that does not migrate during chromatography has a value of $R_F = 0$ and $k = \infty$; for $R_F = 0.5$, $k = 1$. When the solute moves at the solvent front, $R_F = 1$ and $k = 0$. If the ratio of the two phases in the layer is known (V_m can be found from the difference in the masses of the dry and wet layer), then Equation (5) can be modified to yield

$$K = \frac{V_m}{V_s} \cdot \frac{1 - R_F}{R_F} = \frac{A_s}{A_m} \cdot \frac{1 - R_F}{R_F} \tag{7}$$

which can be used to calculate the distribution constant of the given substance in the given chromatographic system.

The real R_F value ($R_{F,real}$) mostly differs from the observed R_F value ($R_{F,obs}$),

$$R_{F,obs} = R_{F,real} \cdot \Theta = \frac{z_i}{z_F - z_o} \tag{8}$$

where z_i is the chromatographic path length of substance i, z_F is the migration distance of the front, z_o is the distance from the start to the surface of the system in the elution chamber and Θ is a factor with a value from 0.8 to 0.9[18]. This difference between the two quantities follows primarily from the fact that the V_m/V_s ratio decreases during movement of the solvent from the start to the front and thus an undesirable volume gradient is formed [21]. This is a result primarily of the faster rise of the mobile phase in the narrower capillaries, so that the real front of the liquid system, filling the whole capillary space (i. e. the narrow and wide capillaries) is somewhat lower than the observed front position. Thus the experimentally determined R_F values are, according to Eq. (8), mostly 10–20 % smaller than the real R_F values. Determination of $R_{F,real}$ also assumes maintenance of standard conditions during preparation of the layer for analysis (especially maintenance of the same relative humidity) and during the development, conditions that virtually cannot be fulfilled in TLC practice.

If the chromatography is carried out in a chamber saturated with vapours of the solvent system — the most common elution conditions employed in practice — then the molecules of the vapours condense in the capillary space of the layer to form a liquid phase with volume V_v. The greater this volume, the less mobile phase that is required to fill the overall volume V_m. Then the flow of the mobile phase is smaller and the path length of the separated zones (z_i) is shorter. Eq. (8) must then be modified:

$$R_{F, obs} = R_{F, real} \cdot \frac{\Theta}{\xi_v} \tag{9}$$

where ξ_v is a correction factor for the decreased flow through the layer as a result of saturation,

$$\xi_v = \frac{V_m}{V_m - V_v} \tag{10}$$

which has values of 1.1–1.6 depending on the type of adsorbent, composition of the elution system, temperature in the chamber and presaturation time[21].

It follows from Eq. (9) that the experimentally determined retardation factor must always be less than unity. For example, for the most common values of correction factors Θ and ξ_v (0.9 and 1.1–1.2, respectively), $R_{F, obs}$ for a component migrating with the solvent front (i. e. $R_{F, real} = 1$) is only 0.82 to 0.75.

The magnitude of the retardation factor is also strongly affected by the shape of the particular adsorption isotherm, describing the adsorption equilibrium in the system,

$$c_s = K \cdot c_m \tag{11}$$

Three types of isotherm are especially important in adsorption chromatography, linear, convex and concave (curvature related to the c_m axis). If the applied amounts of the analysed substance is not greater than 5–50 μg, the isotherm is

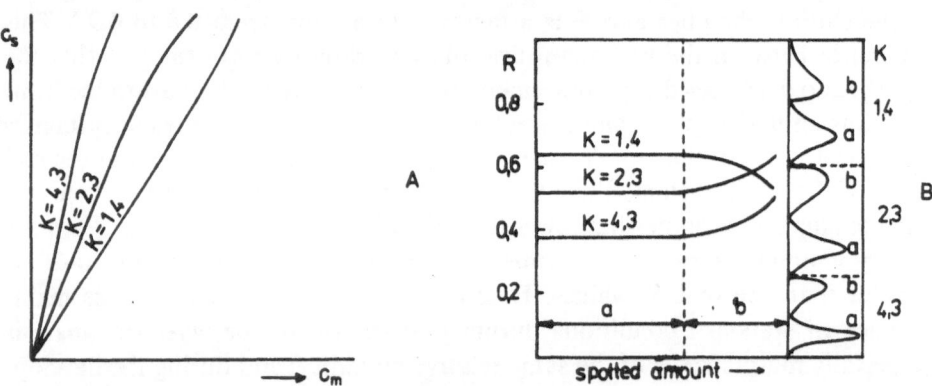

Fig. 1. Dependence of the R_F values and symmetry of the chromatographic zones on the shape of the adsorption isotherm. Constructed using Equation (5) for $A_{st}/A_m = 0.4$. A adsorption isotherms, B retardation factors and shapes of the chromatographic peaks, *a* linear part, *b* non-linear part of the isotherm, R retardation factor, K distribution constant

mostly linear[19]; at greater amounts it is concave or, sometimes, convex. For concave isotherms, the distribution constant decreases with increasing amount of applied substance and thus the retardation factor increases, as follows from Eq. (5). The dependences are opposite for convex systems. The magnitude of K and type of isotherm have a marked effect on the shape and symmetry of the separated zones. If the zones exhibit tailing towards the start, then the distribution usually corresponds to the concave part of the isotherm (Figure 1). Chromatography proceeding under conditions corresponding to a linear isotherm yields Gaussian-shaped zones.

It is thus apparent that the experimentally determined values given in the literature — and also in this book — are difficult to reproduce and thus yield only necessary information on the sequence of the zones on the layer in the given system.

1.1.2 Flow of the Liquid Phase Through the Layer

In contrast to other chromatographic techniques (such as GLC or HPLC), the movement of the mobile phase through the thin layer is a result of the existence of a pressure difference above the curved surface, proportional to the free surface energy and inversely proportional to the capillary radius. Thus the liquid rises in the capillary system until the pressure difference is compensated by hydrostatic pressure.

The velocity of the liquid front is derived from the basic Cameron and Bell relationship

$$z_F^2 = \varkappa t \tag{12}$$

where $\varkappa$ is the flow constant and t is the time at which the front attains distance z_F [21]. Differentiation of Eq. (12) with respect to dt yields the relationship for the front velocity at point z_F,

$$u_F = \frac{\varkappa}{2 z_F} \tag{13}$$

The constant $\varkappa$ is a characteristic quantity for the flow-rate of the solvent for the given type of layer. It is dependent on the solvent parameters and those of the adsorbent.

$$\varkappa = k_1 \frac{r\gamma}{\eta} \cdot \frac{V_m}{V_m - V_v} = \bar{u}_{zF} \cdot z_F \tag{14}$$

where k_1 is a constant, r is the radius of the capillary system, γ and η are the surface tension and viscosity of the mobile phase, respectively, and $\bar{u}_{zF}$ is the average velocity of the front on the z_F pathway. It follows from Eqs. (13) and (14) that the velocity of the front decreases with increasing distance from the start and is directly proportional to the capillary radius in the thin layer and the surface tension of the mobile phase and inversely proportional to the viscosity. In saturated chambers, the front velocity increases with increasing volume of the gaseous phase, V_v, condensed in the thin-layer capillary system. However, if this condensed phase leads to solvation of the sorbent and thus to a decrease in the effective capillary diameter, then the flow velocity decreases even at large V_v values. The anomalously low flow rate of methanol in saturated chambers can be explained in terms of this phenomenon.

1.1.3 The Effect of Diffusion on the Zone Resolution

After application of the sample at the start of the thin layer, a narrow zone with relatively high solute concentration is formed. As soon as the solvent front reaches the start position, molecules of the solute begin to diffuse from areas with higher concentration to those with lower concentration. In chromatography on layers applied to rods with a circular cross-section (Chromarods), if the sample is applied evenly around the whole circumference, then it is sufficient, as with column chromatography, simply to consider diffusion in the direction of flow of the mobile phase and in the opposite direction (longitudinal diffusion). However, if the sample is applied to only part of the layer, then two-dimensional diffusion must be considered, as with flat layers.

The diffusion rate obeys Fick's laws and is proportional to the diffusion coefficient, defined as the amount of substance (mols) that passes in unit time through a cross-section with unit concentration gradient. Typical diffusion coefficients in liquids are of the order of $10^{-5}\,cm^2\,s^{-1}$. A number of empirical relationships have been derived for estimating these coefficients; the relationship derived by Wilke and Pin Chang[20, 23] is quite well known:

$$D_i = 7.4 \cdot 10^{-8}\ (F_m M_m)^{1/2}\ V_i^{0.6} \cdot T/\eta \qquad (15)$$

where M_m is the molecular mass of the mobile phase, T is the absolute temperature, η is the solvent viscosity, V_i is the solute molar volume and F_m is the association factor, characterizing the effect of intermolecular attractive forces; this factor has a value of about 1 in non-polar solvents and 2.6 in water. The above equation is valid for liquids with small to medium-sized molecules and clearly illustrates that the diffusion coefficient depends on the properties of both the solute and the solvent and increases with temperature.

The concentration profile of the diffusion zone (represented by a chromatographic peak) has the shape of a Gaussian curve, with width proportional to twice the product of the square roots of the diffusion coefficient and of the time that the solvent remained in the system, as given by the Einstein equation,

$$\sigma^2 = 2 D_i t \tag{16}$$

where σ is the half-width of the Gaussian curve at the inflection point. At a constant flow-rate of the mobile phase through the layer (which is a condition difficult to realize in TLC) the peak width is proportional to the square root of the distance travelled by the solute in the system.

Assuming that the diffusion coefficient is constant and diffusion occurs only in the direction of flow of the mobile phase (the x-axis in Figure 2), solution of Fick's second law yields an expression for the solute concentration at an arbitrary point x,

$$c = c_{max} \cdot \exp\left(-\frac{x^2}{4 D_i t}\right) \tag{17}$$

and application of Eq. (16) yields

$$c = c_{max} \cdot \exp\left(-\frac{x^2}{2 \sigma^2}\right) \tag{18}$$

where c_{max} is the maximum concentration of the component at the point $x = 0$.

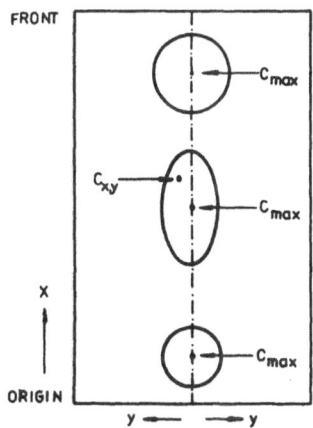

Fig. 2. Idealized shape and size of the chromatographic zones for two-dimensional diffusion in dependence on the retardation factor value. x direction of motion of the mobile phase on Chromarods, the axis of the rod; y direction perpendicular to x on Chromarods, the circumference of the rod, $y = 0$ point of sample application; c_{max} maximum concentration of the solute in the zone; $c_{x,y}$ concentration of solute at the point defined by coordinates x, y; $x = 0$ is the origin

In planar chromatography (and under certain conditions during separations on Chromarod layers), two-dimensional diffusion is involved. The theoretical description then becomes more difficult. The dynamic theory of two-dimensional diffusion on thin layers, developed by Belenkii et al[24], describes the concentration of solute at point x, y (Figure 2) at time t, in terms of an equation for the two-dimensional Gaussian distribution

$$c_{x,y} = c_{max} \cdot \exp\left[-\frac{1}{2}\left(\frac{(x-vt)^2}{\sigma_x^2} + \frac{y^2}{\sigma_y^2} \right) \right]$$
(19)

where v is the velocity of the zone, σ_x and σ_y are the standard deviations characterizing the spot widths in the direction of movement of the zones and in the perpendicular direction, respectively, and c_{max} is the maximal solute concentration at the centre of the zone. Belenkii derived the following relationships for the latter three parameters:

$$\sigma_x^2 = 2(D_i + kv\tau)R_F t$$
(20)

$$\sigma_y^2 = 2D_i R_F t$$
(21)

$$c_{max} = \frac{m_a}{4\pi R_F t[D_i(D_i + kv\tau)]^{1/2}}$$
(22)

where m_a is the amount of sample applied, k is the capacity ratio (see Eq. (6)) and τ is a parameter characterizing the time during which excess diffusion occurs. This value can be estimated from the sorbent grain diameter (dp) and from the diffusion coefficient D_i,

$$\tau \approx \frac{dp^2}{60D_i}$$
(23)

It follows from Eqs. (20) and (21) that both zone widths increase with increasing elution time and R_F value. The shape of the zone — ellipse or circle — depends on the magnitude of the parameter τ and the value of the retardation factor. In layers consisting of very fine sorbent particles (dp $< 5 \times 10^{-4}$ cm), τ approaches zero (at $D_i \sim 10^{-5}$ cm^2 s^{-1}, $\tau < 4 \times 10^{-4}$ s) and $\sigma_x^2 \approx \sigma_y^2$. Similarly, at $R_F \to 0$ or $R_F \to 1$, v or k also approaches zero and it then follows from Eq. (21) that $\sigma_x = \sigma_y$. Hence, the shapes of the zones close to the start and the front are roughly circular and in the middle of the layer ($R_F \sim 0.5$) are elliptical, assuming that the sorbent grain size is large ($> 10^{-3}$ cm).

1.1.4 The Theoretical Plate; Resolution; Separation Number

Comparison of separation efficiencies is still based on the concept of a theoretical plate, introduced into chromatography by Martin and Synge[22] and used in TLC in spite of a number of objections[6,21]. According to this concept, the chromatographic layer consists of a number of hypothetical sections in which equilibrium is established in the distribution of the solute between the mobile and stationary phases. Assuming that the distribution constant value remains constant in all these plates and diffusion in the direction of flow can be neglected, the number of plates is given by the relationship

$$n = n_1 (z_F - z_o) \qquad (24)$$

where n_1 is the number of plates per unit length. Its reciprocal value

$$H = \frac{1}{n_1} = \frac{z_F - z_o}{n} \qquad (25)$$

is the height equivalent to a theoretical plate (HETP). The average plate height can be derived from the zone width and path length z_x

$$\bar{H} = \frac{\sigma_x^2}{z_x} = \frac{b^2}{16 z_x} = \frac{b_{0.5}^2}{5.54 z_x} \qquad (26)$$

where b or $b_{0.5}$ is the width of the Gaussian curve — peak — at the base ($b = 4\sigma$) or at half-height ($b_{0.5} = 2.354\,\sigma$), respectively. In Eq. (26) the height of a theoretical plate is the average value of the heights of all the plates through which the zone passes along path z_x. The overall number of average plates for the whole length of the layer (i.e., 15–20 cm for classical layers, 10 cm for Chromarods, 5–10 cm for HPTLC and 10–20 cm for HPTLC—OPLC) can be calculated from the experimentally determined zone widths (from Eqs. (25) and (26)) assuming that the zone of the given solute migrates over the whole layer up to the front, which is realistic only for flow-through chomatography (readily realizable for OPLC, but not for Chromarods).

In chromatographic processes where the mobile phase migrates through the layer, equilibrium distribution of the solute in both phases can never be established, in contradiction to the concept of the theoretical plate described above. The first coherent theory considering this fact is the rate theory developed by van Deemter et al[25]. This theory states that the zone broadening during elution is a result of several independent factors, eddy diffusion in the mobile phase (term A in Eq. (27)), molecular diffusion (term B) and resistance to mass transfer in the mobile phase (term C):

$$H = H_A + H_B + H_C \qquad (27)$$

Eddy diffusion of the mobile phase is a result of different local velocities of the phase as a result of different dimensions of the capillary system of the layer. The mobile phase flows faster in wider capillaries than in narrower ones and thus the solute molecules progress at different velocities in the layer; some move in a straight line, others according to the solvent flow around the sorbent particles.

Thus term A is proportional to the size of the solid phase particles (dp), their shape and the homogeneity of the whole layer; the following relationship was derived for this dependence[25]:

$$H_A = 2\lambda\,dp \tag{28}$$

where λ is a dimensionless factor, dependent on the geometry of the layer. It can attain values of 1 to 20 in adsorption chromatography[21].

Molecular diffusion in the mobile phase is described by the Einstein law mentioned above and is a result of the concentration gradient between the zone and the mobile phase. If the zone lies across the whole width of the thin layer (line application for flat layers, application to rotating Chromarods), then only one-dimensional longitudinal diffusion need be considered and term B is defined by the relationship

$$H_B = \frac{2\psi D_i}{u} \tag{29}$$

where ψ is a dimensionless factor including the resistance of the layer to free molecular diffusion, generally with a value of 0.6 to 1.

The resistance to mass transfer in the mobile phase is a result of various linear velocities of the mobile phase in the vicinity of the surface of the sorbent particles and in the centre of the capillaries and the consequent differences in the velocities of the solute molecules in the layer. These differences are compensated by diffusion transfer of the solute from one stream to another. The distance travelled by these molecules during such diffusion is dependent on the capillary diameter, which is proportional to the particle size dp. The value of term C can then be calculated from the equation

$$C = \frac{\omega\,dp^2 u}{D_i} \tag{30}$$

where ω is a factor characterizing the type of sorbent, mostly with a value greater than 1. Substitution of Eqs. (28) to (30) into Eq. (27) yields the basic form of the van Deemter equation,

$$H = 2\lambda\,dp + \frac{2\psi D_i}{u} + \frac{\omega\,dp^2 u}{D_i} \tag{31}$$

In addition to this equation, a number of others have been proposed, the most important of which is the relationship introduced by Knox[26] for parameter optimization in liquid chromatography:

$$h = A v^{1/3} + B/v + Cv \tag{32}$$

where h is the reduced plate height H/dp and v is the reduced velocity of the mobile phase,

$$v = u\,dp/D_i \tag{33}$$

The significance of the individual terms in Eq. (32) is the same as in Eq. (27); once again, the term $Av^{1/3}$ corresponds to the contribution of eddy diffusion, where the constant A expresses the quality of the column packing ($A < 1$, homogeneous packing, $A = 2$ to 3, inhomogeneous). Term B/v characterizes the molecular diffusion ($B \approx 2$) and Cv is the contribution arising from the resistance to mass transfer ($C \sim 0.02\text{--}0.2$).

In TLC with low mobile-phase flow-rates, the term B/v, i.e. molecular diffusion, is most important, along with term $Av^{1/3}$ to a certain degree.

However, Eqs. (31) and (32) are not suitable for calculation of the efficiency of thin layers because, as follows from Eq. (13), the velocity u is not constant along the whole pathway z_F and the value of H thus depends on the position of the zone on the chromatogram.

DeLigny[27] modified Eq. (31) for TLC, so that the average height equivalent to a theoretical plate can be expressed as

$$\bar{H} = 2\lambda\,dp + 2\psi D_i \frac{z_F - z_o}{\varkappa} + \frac{\omega\,dp^2}{D_i}\ln\frac{z_F}{z_o} \tag{34}$$

where $\varkappa$ is the flow constant (Eq. (14)), z_F is the distance of the front from the surface of the mobile phase in the chamber and z_o is the distance of the origin from the surface.

The importance of Eq. (34) is limited by the fact that the values of coefficients λ, ψ an ω cannot be found accurately, as they depend markedly on the geometry of the layer, which is difficult to define. However, some interesting conclusions are useful in chromatographic practice. First, the average plate height is proportional to the elution pathway z_F, but not to the position of the zone on the chromatogram, i.e. to R_F. The plate height decreases and separation efficiency increases with increasing distance of the origin from the surface (z_o), which is understandable considering the greater velocity of the mobile phase (and thus slower establishment of equilibrium) in the lower part of the layer. The greater the value of the flow constant $\varkappa$, the less important is diffusion term B; in contrast, the importance of term C is directly proportional to the magnitude of $\varkappa$. Hence, the curve depicting the dependence of the plate height on distance z_F must pass through a minimum. It has been calculated that, for common types of thin layers, this minimum lies about 10 cm from the start (for $z_o = 1$ cm)[21]. Thus the optimal value of the flow constant $\varkappa$ is related to minimum H and can be found by differentiation of Eq. (34) (for $dH/d\varkappa = 0$)

$$\varkappa_{opt} = \frac{z_F D_i}{dp}\left(\frac{2\psi}{\omega\ln\frac{z_F}{z_o}}\right)^{1/2} \tag{35}$$

For slowly flowing solvents, for which $x < x_{opt}$, the B term is most important, i.e. general molecular diffusion; on the other hand, in rapidly flowing systems, the plate height is affected primarily by resistance to mass transfer in the mobile phase (term C)[21]. After substitution of the probable values of parameter ψ (0.6) and ω (1.3) and common values of z_o and z_F (1 cm and 10 cm for Chromarods), Eq. (35) can be used to estimate the optimal flow constants for elution solvents,

$$x_{opt} = \frac{0.63 \cdot D_i}{dp} \tag{36}$$

Thus it is preferable to employ fast-flowing systems for fine layers, and slow-flowing systems for coarser layers. However, the practical applicability of this equation is limited because the values of ψ and ω are not constant and vary with the composition of the solvent.

The efficiency of the separation layer is closely connected with the height equivalent to a theoretical plate, i.e. the ability of the layer to resolve the individual components of the analysed sample. The separation of the components in the layer can be either complete or incomplete; the degree of separation can be quantitatively characterized by the resolution R_s, expressed by the equation

$$R_s = \frac{z_{x2} - z_{x1}}{2(\sigma_2 + \sigma_1)} = \frac{R_{F2} - R_{F1}}{2(\sigma_2 - \sigma_1)} (z_F - z_o) \tag{37}$$

This relationship defines the resolution as the distance between the apex of two neighbouring peaks divided by twice the standard deviation, i.e. the sum of the widths of both zones at 60.7 % of their heights (Figure 3). For a value of

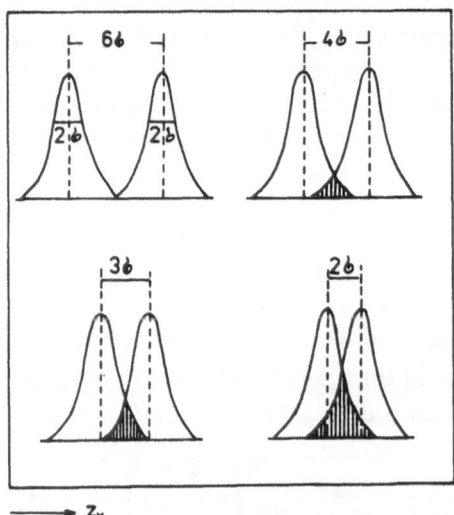

Fig. 3. Zone separation at various resolution values R_s

$R_s = 1.0$, the distance between the apices is 4σ (called 4σ resolution) and the peaks overlap by about 3 %. Other common types of resolution are apparent in Figure 3.

Eq. (37) permits quantitative evaluation of this quantity, but yields no information on the chromatographic parameters that should be selected to attain maximum resolution. Snyder[19] considered this problem and derived an equation expressing the resolution as an explicit function of three terms,

$$R_s = 1/4\left(\frac{K_2}{K_1} - 1\right) \cdot (\bar{R}_F n)^{1/2} \cdot (1 - \bar{R}_F) \tag{38}$$

the first expression in brackets is called the selectivity and describes the effect of the ratio of the distribution constants of the two solutes. The higher the selectivity, the better the zone resolution. The selectivity can be affected by the choice of stationary and mobile phases.

The expression $(\bar{R}_F n)^{1/2}$ describes the effect of the quality of the layer, i. e. of the number of plates n. It should be noted that the resolution increases only proportional to the square root of n; in other words, a ten-fold increase in the number of plates leads to only a three fold increase in the resolution. Another parameter in this term is the average R_F value and when small (the square root is involved), this has a favourable effect on the resolution. At larger R_F values, the negative effect of the third term $(1 - \bar{R}_F)$ begins to appear. The contradictory effect of the retardation factor appears as a maximum on the curve expressing the dependence of the R_s value on R_F. It can be seen from Figure 4 that the maximum resolution region lies between R_F values of 0.3 and 0.4, irrespective of

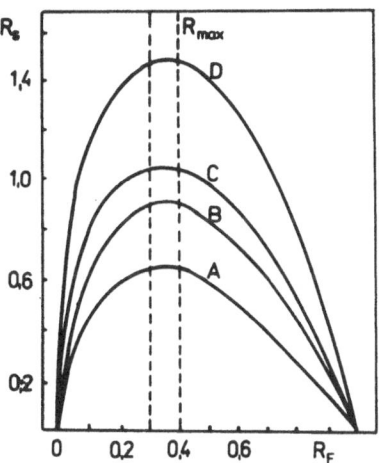

Fig. 4. Dependence of the resolution R_s on the average R_F value for a separated pair: A-for $K_2/K_1 = 1.3$, $n = 500$; B-for $K_2/K_1 = 1.5$, $n = 1\,000$; C-for $K_2/K_1 = 1.3$, $n = 1\,000$; D-for $K_2/K_1 = 1.5$, n = 1\,000$. Calculated using Eq. (38)

the selectivity value and the number of plates. This fact is well known in chromatographic practice and appears primarily in the separation of critical pairs (e. g. positional isomers), with similar distribution constants.

A further practical quantity characterizing the efficiency of the layer in a concrete chromatographic system is the separation number (SN), defined as the number of zones that are completely separated (i. e. with a resolution of 4.7σ) between $R_F = 0$ and $R_F = 1$. Kaiser[6,28] states that this quantity can be calculated from the equation

$$SN = \frac{z_F - z_o}{b_s + b_F} - 1 \tag{39}$$

where b_s and b_F are the peak widths at the start and front, respectively, measured at the peak half-height. Both values can be found by extrapolating a function expressing the dependence of $b_{0.5}$ of the peak on its R_F value. In isocratic chromatography, i. e. in chromatography without a gradient in the composition and flow-rate of the mobile phase, this function is a straight line that, when extrapolated to the start and front, yields b_s and b_F.

1.2 Methods of Thin-Layer Chromatography with Flame Ionization Detection (TLC-FID)

This chapter deals with the techniques of chromatography performed on Chromarods and detection of the separated zones using an Iatroscan TH-10 Analyzer. Basic data is provided on the apparatus and accessories available at the time of writing. The methodological part contains a detailed description of verified procedures and gives information on the advantages, drawbacks and limitations of TLC-FID application.

1.2.1 A Short Description of the TLC-FID Procedure

Chromarods (10 pieces) are placed in the rod holder (Figure 5), cleaned and activated in the flame of the detector of Iatroscan TH-10 instrument. The holder is removed from the instrument and placed on the spotting guide. An amount of 1–50 µg of sample in 0.2 to 2 µl of solution is applied to the start using a micropipette. The solution is applied in several aliquots so that the band of applied sample is not wider than 2–3 mm. The thin layers are eluted in the development tank, usually lined with a wick of chromatographic paper, and dried prior to detection for 3–10 min in a small drying oven at 60–120 °C, depending on the volatility of the solvent system used. Then the rods are placed

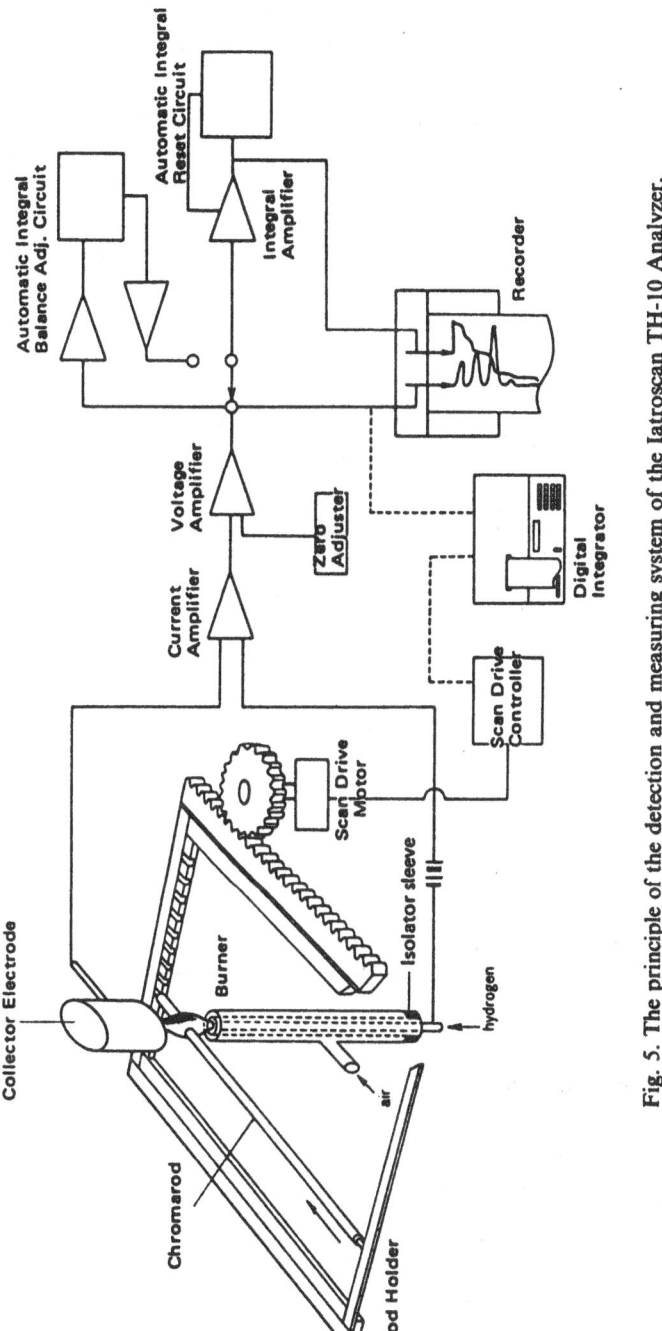

Fig. 5. The principle of the detection and measuring system of the Iatroscan TH-10 Analyzer, Mark III

in the sliding frame of the Iatroscan TH-10 Analyzer and passed through an FID. The individual separated zones are ionized in a hydrogen flame and the ionization current produced is amplified and fed into the integrator and recorder.

The detection principle and measuring system are also shown in Figure 5.

The thin layer is activated by passing it through a hydrogen flame which prepares it for the next analysis without any other treatment.

Chromarods cannot, of course, be used for two-dimensional chromatography. This disadvantage is partly offset by selective scanning of the separated components during which the thin layer is first eluted in a less polar solvent system and the instrument detects the components separated between R_F 0.1 and R_F 1.0. The polar substances that remain at the start are then separated in a more polar system.

Multiple elution in one or several solvent systems can readily be carried out; selection of a suitable length of the elution pathway and solvent polarity permits very good separation of quite complex mixtures.

Chromarods can also be impregnated with silver nitrate or boric acid to improve the resolution of some critical pairs (separation on the basis of the number of double bonds, separation of positional isomers, etc.).

1.2.2 The TLC-FID Technique

The apparatus for TLC-FID comprises:
 – a system detection in an FID
 – a recorder or digital integrator with a plotter
 – accessories for sample application, elution and drying of the Chromarods.
These are described in the following three sections.

1.2.2.1 The System for Detection in the FID

At the beginning of the 1970s the Japanese company Iatron Inc. introduced Thinchrograph TFG 10, which was soon replaced by the more modern Iatroscan TH-10 Analyzer, Mark II, which is widely used in Japan, the USA, Canada and Europe.

In the 1980s the Mark III version appeared, differing from the earlier version in the regulation of the scanning rate and the manner of placing the Chromarods in the detector. The procedure given below is intended for this type of instrument and differences in the construction of the Mark II and Mark III Analyzers are mentioned at the end of this section.

The Iatroscan TH-10 Analyzer Mark III is an automatic system, with the following main parts:

A. **The detection part** (Figure 6) consisting of a flame-ionization detector with a burner (1) and a collector electrode (2), along with a moveable frame for fixing the Chromarod holder (3) (the rod holder locating frame), angled at 30° to the axis of the flame and permitting constant movement of the rod in the FID. The frame is fitted with notches for individual Chromarods if they are to be located individually in the scanner.

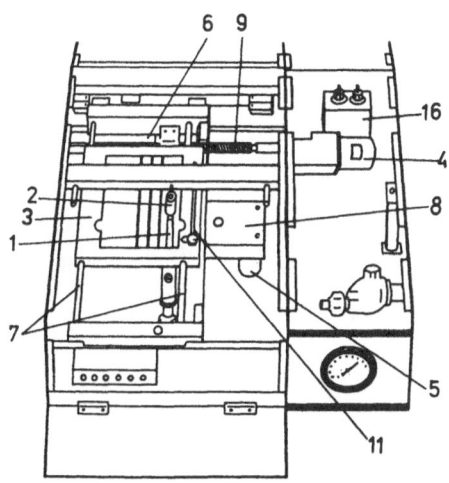

Fig. 6. Interior of TH-10 Mark III series showing principal parts. (1) H_2 burner; (2) FID collector electrode; (3) rod holder; (4), (5) drive motors; (6) lead screw; (7), (9) guide rods; (8) screw drive; (11) peak pyrolysis selector; (16) air filter

The burner (negative pole of the FID) has three jets for introduction of hydrogen (160–180 ml min^{-1}), lying perpendicular to the Chromarod axis, similar to the AAS flame profile. A rather strong stream of air (2.0–2.2 l min^{-1}) is passed trough the gap between the mantle and the central part of the burner to prevent contamination by dust particles from the atmosphere in the hydrogen flame and the effectively contribute to the stability of the background noise level of the detector. The collector electrode (the positive pole of the FID) is shaped like a hollow cylinder and fixed to an inclinable top cover.

The rod should lie exactly in the centre of the hydrogen flame and the distance between the surface of the layer and the surface of the nozzle should be 0.3–0.5 mm.

B. **The moving part** (Figure 6) consists of two stepping motors (4 and 5), the first of which drives a lead screw (6) to control horizontal shift of the frame (3) along the metal guide rods (7). The second motor with a gear drive (8) permits vertical movement of the frame.

The required scanning speed of the rods in the detector can be selected by adjusting the rotation rate of the motor (5) using the scan selector switch (10, Figure 7). The scan rates available for Mark II and Mark III models are given in Table 1. Generally rates 2 to 4 (Mark III) are selected, so that detection of 10 rods takes only 10 to 15 min. In contrast to Mark III, the scanning speed of the rod holder in Mark II can be readily adjusted by replacing one of the gears in the gear drive.

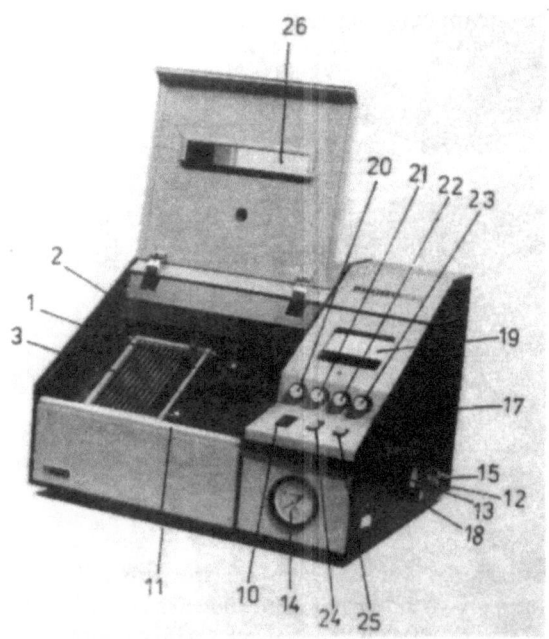

Fig. 7. The Iatroscan TH-10 Mark III with cover open. (1)-(3), (11) see Fig. 6; (10) scan selector swich; (12) hydrogen inlet connector; (13) hydrogen pressure control knob; (14) hydrogen pressure gauge; (15) air inlet connector; (17) air flow meter; (18) air control knob; (19) ammeter; (20) start and reset switch; (21) return switch; (22) auto zero switch, (23) zero adjust knob; (24) pause switch; (25) auxiliary switch; (26) observation window

Table 1 Adjustable Velocity of the Chromarods in the FID

Mark II			Mark III		
Gear no.	Scan. speed cm s^{-1}	Sec/rod scan	Speed selector no.	Scan. speed cm s^{-1}	Sec/rod scan
70	0.73	17	5	0.51	25
55	0.58	22	4	0.42	30
40	0.42	30	3	0.36	35
30	0.31	40	2	0.31	40
20	0.21	60	1	0.25	50
12	0.13	92	–	–	–

Chromarods are usually scanned in the flame from the front to the start. After completion of detection of the first rod, the rod is indexed horizontally to half the distance to the next rod and then returns automatically to the FID-scan position for the next rod. The range of vertical movement of the frame and thus the range of detection can be adjusted using the peak pyrolysis selector (11, Figure 6). This function can be used to program the combustion of selected components on the rod, leaving those remaining to be further chromatographically developed and scanned. The detection and moving parts are covered by a lid with an observation window, permitting observation of the progress of the scanning sequence.

C. **The gas system** (Figure 7). Hydrogen enters the instrument through inlet (12) and passes through the regulating valve (13) and pressure gauge (14) into the lower part of the burner (1, Figure 6). Air from the pump (not depicted) or a pressure cylinder is fed in through inlet (15) and through filter (16), packed with activated carbon, into the flow-meter (17) (0–3 1 min^{-1}) with regulator (18) and then into the central part of the burner. The purity of the hydrogen and also of the air is a necessary precondition for low detector noise. Thus, air from a high-pressure cylinder is used, as organic solvent vapours or cigarette smoke in the laboratory air can lead to considerable variation in the FID baseline.

D. **The electronic and control part** consists of a system of microswitches and relays that automatically controls the horizontal and vertical movement of the frame and also the scanning of the individual rods.

The next part is a differential and integrating FID amplifier. Both these signals may then pass to a double-channel recorder or alternatively the analogue signal is amplified and passed through a suitable digital integrator to a single-channel recorder or plotter. This part also contains a microammeter (19), used to observe the output signal during thin layer clean-up and activation.

The Iatroscan TH-10 Mark III instrument is switched on by "start and reset" button (20, Figure 7); and the scanning operation can be interrupted by the "return" switch (21). The moveable rod holder then returns to the initial position when the reset button (20) is pressed.

The "auto-zero" switch (22) is used for automatic calibration of the baseline signal. The "zero-adjust" potentiometer (23) is employed to adjust the zero value on the microammeter (19). The "pause" rocker switch (24) in the "on" position interrupts the automated instrument function until the digital integrator finishes treating and printing out the data from the detection of the previous rod. After switching to the "off" position, the scanning sequence is automatically restarted.

The additional rocker "aux switch" (25) in the "on" position synchronizes the scanning rythm with the function of the externally linked electronic integrator and sets the blank scan time, i. e. the time between the scanning of two subsequent rods, to 30 s. This is sufficient for complete data treatment and print-out of the results. When using a two-channel recorder without a digital integrator, the "aux" switch is in the "off" position and the blank scan time is reset to 10 s, decreasing the time required for detecting 10 Chromarods by more than 3 min.

The electronic part also includes an input voltage selector, a socket for connection to the mains and sockets for the air pump, recorder, electronic integrator and instrument grounding.

The Iatroscan TH-10 Mark II differs from Mark III in the following features:

– It is not equipped with a dual-purpose development and scanning rod holder and thus the individual Chromarods must be inserted using a pair of tweezers. This is more tedious and the rods are easily broken or damaged. A universal holder was later constructed for Mark II, but requires modification of the instrument and symmetrical placing in the detector, which proved rather difficult.

– The detection rate cannot be adjusted using an electronic selector, but must be varied by exchanging gears.

– Model II does not have "pause" (24) and "aux" (25) switches (Figure 7) and thus the blank scan time cannot be varied (this is dependent on the gear used).

– The zero line is not calibrated automatically.

– The collector electrode is not pivoted and must be tediously removed during every adjustment of the distance of the Chromarod from the burner nozzle.

– The top cover does not have a window.

1.2.2.2 Recorders and Electronic Integrators

A recorder is used in the treatment of the FID signal, either a single-channel recorder connected to an electronic integrator, or a two-channel recorder, where one channel is used for graphical recording of the time dependence of the detector response, i.e. the actual chromatogram, and the second for recording of the integration traces. The recorder should have a measuring range of 10 mV to 1 V (measurements are usually carried out in the range 50–200 mV) and the chart speed should be 50–500 mm min^{-1} (usually 100–200 mm min^{-1}). Most line recorders used in GLC and HPLC satisfy these requirements.

At present, TLC-FID chromatograms, similar to GLC and HPLC, are increasingly evaluated using electronic integration instruments containing programs for separation of resolved and incompletely resolved double peaks and groups of peaks, for calculation of corrected peak areas, according to the FID response of the given components. In most cases, normalization programs are included for calculating the composition of the analysed mixture, either by comparison with a standard calibration mixture, or by the method of internal or external standards. System 1 is such an instrument, as is also, to a certain degree, the Minigrator instrument; both of these were manufactured by Spectra Physics.

Newer types of instrument are equipped with a plotter (Hewlett Packard Reporting Integrator 3390 A, Intelligent Integrator 7000 A SIC, Japan), so that a recorder is no longer necessary. The SP 4200 Computing Integrator (Spectra Physics) is a useful instrument for linking to the Iatroscan TH-10 Analyzer; the SP 4270 model can also be used. These instruments print out chromatograms and calibration curves, calculated by the least squares method and yield analytical information including statistical evaluation of the results. In addition, they automatically adjust optimal parameters for evaluation of the individual peaks, eliminating problems with programming of optimal peak widths.

The accuracy of integration of detector signals is determined by programmable parameters in quite a broad range; these include parameters determining the sensitivity of peak detection (slope sensitivity), manner of detecting groups of peaks, detection of tailing peaks, etc. Thus the reliability of the results depends on the quality of the evaluated chromatogram and the quality of the integrator, as well as on the experience and of the operator and the accuracy of the logical evaluation.

A critical drawback of most electronic integrators is the manner in which the base line is reconstructed. If this parameter fluctuates during the measurement, which can happen in TLC-FID (because of the effect of non-uniformity of the layer on the detector noise), then its incorrect localization can substantially affect the areas determined, especially for small peaks. Then the evaluation of the chromatograms by classical manual methods (triangulation, weighing, etc.) is sometimes more accurate than integrator calculation.

In spite of their rather high price (comparable with that of the Iatroscan TH-10), electronic digital integrators are single-purpose instruments that can treat the analogue signal from only a single detector at any given time and yield only a list of the areas, concentrations and retention times of the individual peaks on the chromatogram. This data must then be further treated mathematically to yield an overall analytical report.

These problems can basically be solved in two ways: in the simpler approach the integrator is combined with a cheaper type of personal or laboratory microcomputer (e. g. the Apple II). This hybrid system is used, for example, in the quantitative evaluation of HPTLC chromatograms[29]; the integrator measures the areas of the individual peaks and feeds then through a serial interface to the computer which stores this information on floppy disks and computes a report which is then printed out or displayed on a screen.

Units combining the advantages of more complex electronic integrators with flexible processor systems represent a further stage in instrumental development. They can simultaneously collect raw data from a number of detectors, store this

information in random access memory (with a capacity of 160 or more kiloby-
tes), where the memory registers are employed not only for the actual on-line
chromatographic evaluation (i.e. collection and evaluation of the detector
information), but also for calculations and operations using external data series.
Here, it is important that the computer works in a shared regime, with preferential handling of the detector signal. The primary data and adjusted values are
stored in the peripheral disk memories and evaluated after completion of the
detection cycle. Thus conditions are optimal for evaluation of the whole chromatogram and also for finding the optimal zero line position and satisfactory
evaluation of very complex chromatograms with a number of overlapping
peaks. Such systems (e. g. PU 4850 manufactured by Philips) even permit
subtraction of two different chromatograms and depiction and subsequent
evaluation of the differential recordings. This principle can be used to increase
the sensitivity of TLC-FID by storing the "noise profile" of the individual
Iatroscan rods in the memory during the first scan, and then subtracting
this value from the actual chromatogram of the analysed mixture. Chromatograms of more important samples, e. g. standard mixtures, can be converted to bar graph profiles using a suitable program. These can then be stored in
disk memories. The stored chromatograms can be compared on the display
screen for normalization of retention times and printed out if necessary. The
analytical results are printed out according to the given program as a laboratory
report (including statistical evaluation, normalization, graphical recording of
chromatograms and of graphs of linear and nonlinear dependences of the
selected variables). The higher price of these systems is compensated for by their
extensive flexibility and the possibility of simultaneous treatment of data from
a number of analysers, including GLC and HPLC. These instruments will find
even greater application in TLC-FID after completion of the development of
new types of detectors (e. g. FID with simultaneous use of FTID), whose
analogue signals can be simultaneously processed and compared.

1.2.2.3 Accessories for Application, Elution and Drying of Thin Layers

Samples are at present most often applied using microlitre syringes with 0.1 µl
divisions. Among the available syringes, the following are especially useful:
Hamilton (Switzerland), Gazochrom (U.S.S.R) or Scientific Glass Engineering
(Australia, England). As the width of the applied trace should be as small as
possible, a small amount of the most concentrated solution possible should be
applied. Then capillary spotting tubes are useful, e. g. the type used in HPTLC[29]
(a platinum-iridium capillary with dosage volumes of 100, 200 or 500 µl, manufactured by Antech GmbH, Bad Dürkheim, FRG). The Newman-Howells
company sells an automatic applicator for spotting of a more uniform distribu-

tion of the sample solution around the circumference of the Chromarods. This device applies 1–50 μl of sample to each of 10 rods rotating during the application. The sample band width is very narrow (less than 1 mm), which results in better resolution and improved reproducibility of the analysis.

Application at the origin point of the rods is facilitated by the Iatron spotting guide.

The developing chambers are similar to N-chambers for the elution of classical plate thin layers, with dimensions corresponding to the frame holding the set of 10 rods. A smaller type used for the rod holder in the Iatroscan Mark II system has dimensions of $12 \times 2 \times 18$ cm and is filled with 50 ml of solvent; the larger type with dimensions of $14 \times 4 \times 17$ cm is suitable for the Mark III holder and is filled with 70 ml of the solvent system. The Iatron suspended development tank (DT 250) is used for presaturation chromatography. It consists of an external glass chamber ($14.5 \times 4.5 \times 20$ cm) with a glass lid fitted with a frame that can be used to adjust the level of the rod holder above the eluent (Figure 8). The chamber is filled with 100–120 ml of saturating solvent (that may be but need not be identical with the elution system) and is lined with filter paper. The elution system (40–45 ml) is placed in a small stainless steel reservoir into which the Chromarod holder is lowered after completion of the presaturation step. During elution in this type of chamber, the rods are not contaminated by impurities extracted from the filter paper into the saturating solvent system.

Fig. 8. Suspended development tank with Chromarod holder in raised position

A stabilizing aluminium stand and stainless steel spring clamps are provided with the chambers.

The position of the front of the mobile phase can be easily followed during the elution by back-lighting with a suitable cold lamp; Iatron provides a special lamp with a filter, the Iatron rod viewer, for this purpose (Figure 9).

Fig. 9. Rod viewer positioned behind the TLC development tank with shade in lowered position

A special rod drier is used to remove the solvent from the eluted layers (or water from washed layers) (Figure 10). Common laboratory driers are too large and dust that is easily adsorbed on the layers produces spurious signals in the FID detector, affecting the chromatographic trace. However, a special drier can easily be made from a glass beaker (height ca. 20 cm, diameter 12 cm) with fused tubes for introduction of nitrogen (or pure air) and for a thermometer. The beaker is closed by a hinged lid and screwed to a metal stand. It is heated by a 250 W infrared lamp whose distance from the surface of the beaker can be adjusted to regulate the temperature in the drier. The rod holders are placed on a stainless steel grid made of ca. 2 mm wires.

To prevent contamination of the rods by dust or exhalations prior to detection, they are stored in a clean dessicator. The rods are cleaned and regenerated in acids in a large chamber with a glass lid, e. g. a chamber for classical planar layers. The rods are usually washed in water using a large Petri dish with a lid.

1.2.2.4 Chromarods

Laboratory manufacture of thin layers on thin quartz rods requires considerable manual skill and is not recommended. Layers prepared in this way are not sufficiently homogeneous and, in addition, have low resistance to mechanical damage. Fixing of the layers with organic binders is impossible and addition of gypsum to the adsorbent forms sulphates or calcium salts that inhibit separation and increase the detector noise above an acceptable level.

Chromarod thin layers manufactured according to the patent of Okumura and Kadano[15, 16] have a uniform layer thickness, are resistant to acids and weak bases and can be used up to 100 times without a substantial change in their separation ability. A thin layer formed of a suitable sintered mixture of glass powder and adsorbent is coated onto a quartz rod with a diameter of 0.9 mm and a length of 15 cm. Twelve cm of the layer can be used for chromatography (Figure 11). One example of suitable material for the rod is silicate glass Vycor (96 % silicon dioxide with a specific surface area of $150 \, m^2 \, g^{-1}$, manufactured by Corning Glass Works). The powdered glass can be obtained by crushing silicate, borosilicate, sodium or lead glass, or crystallized ceramic glass, in a ball mill.

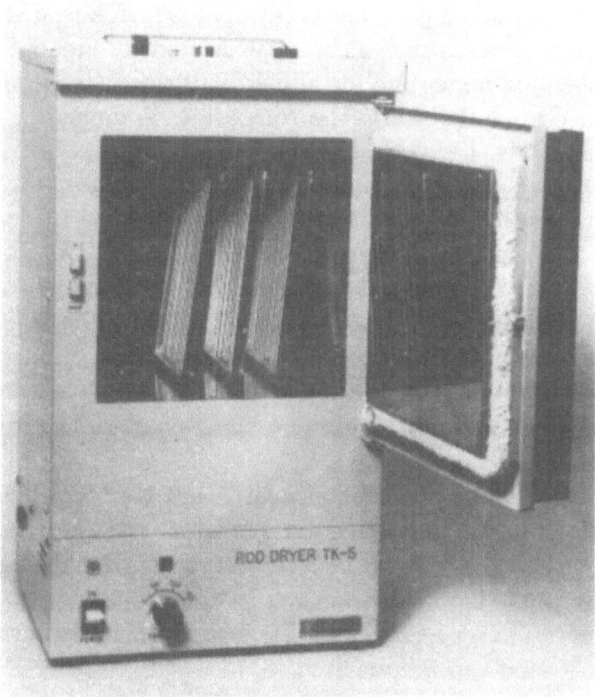

Fig. 10. Rod dryer TK-5 with three Chromarod frames

The separated powdered glass fraction (1 — 10 μm) is mixed with the adsorbent particles (5 — 10 μm) in a ratio of 2 : 1 to 10 : 1. At lower ground glass contents, the layers are too fragile. On the other hand, when the ratio of powdered glass to adsorbent is greater than 10 : 1, the separation ability of the

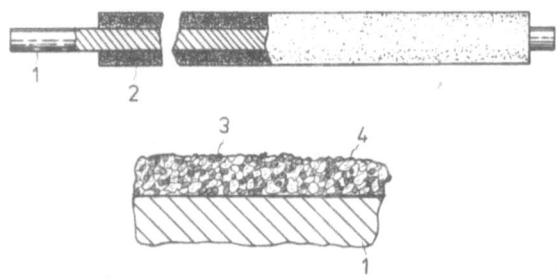

Fig. 11. Cross-section of a Chromarod. (1) quarz rod; (2) thin layer; (3) adsorbent particle; (4) glass particle

thin layer decreases and the R_F values of the separated substances increase disproportionately. In general, layers of high glass content are more suitable for separation of mixtures of polar substances[16] and vice versa.

The mixture of powdered glass and adsorbent is then kneaded after addition of water or organic solvents such as benzene, acetone or ethanol and of a small amount of a substance increasing the adhesion of the suspension to the surface of the rod (e. g. Canada balsam). The rods are then dipped in the slurry and pulled out immediately. The coated rods are dried and baked in an electric oven for 3–10 min at a temperature slightly higher than the melting point of the powdered glass used. A temperature of 570 °C is sufficient for lead glass, while a temperature of up to 900 °C must be used for ceramic glass[15].

If the temperature is too high, the whole layers melts and loses its separation ability. On the other hand, at lower temperatures no sintered layer is formed. The baked thin layer is gradually cooled to avoid the danger of strain caused by heating among the component particles and between the particles and the core rod. This gradual cooling may be omitted when quartz or high-silicate glass is employed[16].

It has been found using an electron microscope that the thin layer forms a spongy structure of partially sintered glass in which the adsorbent particles are fixed (Figure 12). The thickness of the layer can be regulated by varying the ratio of the powdered and liquid components in the suspension; a more viscous suspension yields a thicker layer, and a less viscous a thinner layer. Commonly, layers of 10 to 200 μm can be prepared[15].

Fig. 12. Section of the Chromarod thin layer under an scanning electron microscope. (a) Chromarod A; (b) Chromarod S; (c) Chromarod S II (× 500)

Commercial Chromarods with layer thicknesses of ca. 50–100 μm are sold in cartons of 10. They are protected from damage by soft polyurethane foam padding. The layer becomes contaminated at points of contact with plastics, resulting in a strong signal on passage of the rod through the hydrogen flame of the detector. However, adhering impurities can be readily removed by double scanning of the rods in the FID.

At present, three types of rods are sold:
- Chromarod S — the active sorbent is silica gel, 10 μm,
- Chromarod S II — silica gel 5μm,
- Chromarod A — aluminium oxide 10 μm.

Chromarods S II have higher separation ability than the S type, but are more expensive. The aluminium oxide layers have increased ability to bind water by coordination bonds and are thus suitable for separation of less polar substances, differing either in the molecular shape or type of functional group. As is common in classical TLC, aluminium oxide quite readily forms chelate bonds with compounds forming intramolecular hydrogen bridges, such as hydroxy-acids, glycols, aminoalcohols, etc. These compounds are then more strongly bonded to the adsorbent and have lower R_F values than on silica gel Chromarods A have been found especially useful for separation of stereoisomers, some hormones, steviosides (plant sweeteners), and genins, and for use with solvent systems containing chloroform and acetone.

Table 2 lists the values of some basic Chromarod parameters found experimentally. These values are only informative, as they represent the average values measured for a small set of Chromarods.

Table 2 Basic Parameters of Chromarods[36]

Parameter	Chromarods		
	S	S II	A
Layer thickness (μm)	104 ± 2	54 ± 9	33 ± 6
V_t (mm³)	38,7	19.8	11.6
V_s (mm³)	12.5	7.1	5.5
V_m (mm³)	26.2	12.7	5.1
V_s/V_m	0.48	0.56	1.08

V_t total volume of the layer (the layer is 12 cm long);
V_s volume of the stationary phase;
V_m volume of the mobile phase;
V_s/V_m ratio of the volumes of the stationary and mobile
phases – equals their cross-sections – see Eq. (3).

Special types of Chromarods have been prepared for research purposes, from a mixture of silica gel and florisil (9 : 1), silica gel and aluminium oxide (1 : 1) and florisil. Lipids with the same chain length but different degrees of saturation can be separated using Chromarods S and S II, impregnated with 5 to 25 % silver nitrate and subsequently activated at 120–180 °C for 2 to 5 h[30]. Rods impregnated with silver nitrate are simply dried prior to sample application, as regeneration in the FID would lead to decomposition of the impregnating agent. Acylglycerol isomers can be separated after rod impregnation with boric acid[31]. Oxalic acid impregnated Chromarods S II gave better resolution of phospolipids. A concentration of 0.25 M oxalic acid in acetonitril, with 15 min of impregnation, followed by 60 min of activation at 110°C, provided the ideal conditions for coating [32]. One of the drawbacks of impregnation is that the lifetime of the impregnated layers is reduced.

Chromarods are quite resistant to chemical solutions and solvents, but are easily damaged mechanically, especially by contact with a sharp instruments such as a micropipette needle, tweezers, etc. Although loss of a small section of the adsorbent does not result in a substantial change in the separation ability of the layer, it can lead to a decrease in the reproducibility of the determination.

Originally, the manufacturer recommended storing the rods in a humidified chromatographic chamber lined with filter paper saturated with distilled water. However, it is now considered preferable to store the rods in 9N sulphuric acid or even in chromic acid cleaning mixture[33]. We have found that, after every four or five analyses the rods should be placed in 9N sulphuric acid for at least 1 h and then washed six times in ca. 300 ml distilled water. Approximately once every three days the rods should be freed of mineral salts (e. g. calcium sulphate) by immersion in 30 % hydrochloric acid. Rods maintained in this way have only a very slightly increased noise level with age and can be used for 150 or more analyses if they are not mechanically damaged.

If the Chromarods cannot be cleaned in the above manner to remove contaminants caught in the micropore structure of the sintered layer, the manufacturer recommends that the whole set of rods should first be chromatographed in a chamber containing 60 % sulphuric acid up to the top edge of the sorbent and then immersed individually into the same acid solution in tall round-bottomed test tubes. The rods are then placed in a low-power ultrasonic bath for 15 min, washed thoroughly in distilled water, placed in a holder and stored in a chamber with 32 % relative humidity. This procedure has the disadvantage of being rather tedious and there is a certain danger of damage to the rods in the test tubes. It can be thus recommended only in exceptional circumstances when contaminants causing interference in the detector cannot be removed by repeated washing in concentrated sulphuric acid and dilute hydrochloric or nitric acid.

1.2.3 Operational Parameters and Procedures

In spite of the similarity, in principle, between classical TLC and ordinary TLC-FID, the methods are very different. Thus experience with ordinary TLC has very limited application to TLC-FID and a number of new procedures must be learned not only in the conditioning of the layers for analysis and their maintenance, but also in elution and detection. These procedures, based on the experience of other analysts and the results of our own work, are summarized in the rest of this chapter. The validity of the procedures and recommendations given is, of course, limited, because TLC-FID is a relatively new, developing method.

1.2.3.1 Preparation of Chromarods for Analysis

Prior to the first analysis, it is necessary to remove impurities adsorbed from the polyurethane foam packing material by double pyrolysis in the Iatroscan instrument under ordinary scanning conditions (see below).

Although the dimensions of all the rods and the thickness of the sintered adsorbent are practically identical, the chromatographic properties of the individual sets and rods are not always the same. Some authors recommend matching Chromarods in sets of 10 with similar flow-rate characteristics, determined using a single solvent system such as diethyl ether[34, 35]. It is claimed that these selected sets yield more reproducible results with lower coefficients of variation. Selection is sometimes even understood to represent separating the rods into sets on the basis of the averaged results of up to 10 quantitative analyses of model mixtures, e. g. triacylglycerol, fatty acids, cholesterol and cholesterol esters.

Such procedures are, however, not ideal. They are rather tedious, shorten the lifetimes of these quite expensive materials and still do not ensure that the properties of the individual rods will not change with a growing number of analyses. The flow-rate in the thin layer depends on the composition of the solvent system so that, for example selection in diethyl ether will not be satisfactory for chloroform or methanol type solvents, and vice versa. In addition, classification of rods according to the results of quantitative analyses of a single model mixture need not yield results for all types of separated mixtures of different substances. Thus, we have found that it is preferable not to carry out selection and to maintain the rods in a single set together in a single holder until they lose their separating ability, i. e. expose them all to the same effects during elution, detection and storage. If one of the rods has very different properties, then it is preferable to remove it from the set, leaving only eight or nine rods in situ. The removed rods can then be used for other purposes, e. g. for preliminary qualitative selection of a new elution system or for testing rod modification (e. g. using boric acid or silver nitrate), etc.

The rods that have been cleaned and activated in the flame of the FID are then ready for application of the analytical samples. Their purity can be evaluated either by visual observation of the Iatroscan microammeter scale, which should not deviate from zero during the pyrolysis by more than a few several tenths of a millimetre, or by observing the peak areas recorded by the integrator. These areas increase slightly with the ageing of the rods, i. e. with the number of analyses (Table 3).

Table 3 Relative FID Responses* of Clean Chromarods S II[36]

Rod no.	After 2 analyses		After 150 analyses		
	1st activation	2nd activation	1st activation	2nd activation	2nd activation HCl
1	9620	667	65609	1612	257
2	8692	480	25131	725	305
3	3627	280	23013	1039	470
4	8507	356	28101	228	280
5	5369	252	4781	607	567
6	17239	450	12767	201	261
7	3581	230	18575	1310	413
8	8500	720	19401	1098	426
9	10450	230	54528	1502	313
10	16800	250	–	–	–
average	9238	391	27990	925	365
C V %	49	44	66	53	28

* Expressed as the average area obtained from the Spectra Physics System 1 integrator (peak width 1 s, minimum area 200). FID regime: 180 ml H_2 min^{-1}; 2.1 l air min^{-1}; rod velocity in the FID 0.42 cm s^{-1}.

Large responses during the first activation in the FID are due to inadequate removal of water during drying of the washed rods. The actual "cleanliness" of the rods can be evaluated on the basis of the area values of the background noise measured during the second activation, where a certain difference can be observed between new and used rods. The higher noise level of the older rods is apparently connected with the presence of salts that are insoluble in sulphuric acid, which is mostly used for the regeneration and storing of rods overnight. This suggestion is supported by the marked decrease in the noise level after washing the rods for one hour in 20 % hydrochloric acid (see the last column of Table 3).

The background noise level of carefully maintained rods has a negligible effect on the reproducibility of the analysis. Under normal conditions, the width of the separated peaks is about 1 s, which, for 30 s detection time for the whole rod scan, represents only 10–30 relative units (see relative responses in Table 3). This area corresponds to the response to 0.001–0.003 µg of substance, i.e. a fraction of the applied amount (1–100 µg) (see Table 24).

Each set of rods should be designated by a basic identification number (a two-digit number from 01 to 99), followed by the number of the analysis which, considering the lifetime of the rods, should be a three-digit number (001 to 300). The total identification numbers then have five digits.

1.2.3.2 Application of the Analysed Mixture to the Rods

Although the Iatroscan system is even capable of identifying hundredths of a microgram of organic substances, quantitative analyses require at least one microgram of each component for reproducibility. The amount of the mixture applied depends on the number of components, with assumed contents of minority substances, and the concentration range in which the dependence of the FID response is roughly linearly dependent of the amount of substance. For example, a mixture with three or four components where one represents 1 % or less is generally applied in an amount of 100 µg or more at the origin, while a multicomponent mixture is usually not applied in amounts exceeding 20–50 µg as otherwise the separation might be incomplete.

If a sufficiently concentrated solution can be prepared (3 % or more) without the danger that a temperature drop could lead to precipitation of one of the dissolved components, then the solution should be applied in a single aliquot (about 0.3–0.5 µl). More dilute solutions (0.5–1 %) are applied in three to five aliquots of 0.2–0.4 µl so that the width of the start is not greater than 2–3 mm. This bandwidth depends not only on the magnitude of the volume applied, but also on the type of chromatographic layer and solvent characteristics. The applied solutions spread less into silica gel rods (Chromarods S and S II) than on aluminium oxide layers (Chromarods A), where considerable skill is necessary to attain the above width limit.

Using the automatic applicator, it is possible to attain bandwidths of less than 1 mm and thus to increase the utilization of the separation ability of the rods considerably (see Chapter 1.1, Eq. (37) for resolution values, R_s, and Eq. (39) for the separation number SN).

In general, the most advantageous solvents are those that have a sufficiently high boiling point (about 60 °C), so that they do not evaporate at laboratory temperature at the needle tip prior to application to the rod. Conversely, the boiling point should be sufficiently low (below about 80 °C) so that it vaporizes sufficiently rapidly after application without the need to use a drying lamp or hot plate. We have found that these requirements are best fulfilled by acetone, chloroform, methanol, n-hexane, diisopropyl ether, ethyl acetate, trichloroethane, benzene, cyclohexane and trichloromethylene. Sometimes, a small amount of water should be added to a combination of these solvents. Particularly unsuitable solvents include diethyl ether, n-pentane and methylene chloride and, of course, high-boiling-point solvents such as dimethylformamide or dimethylsulphoxide.

During the sample application, the rods are located in a holder placed on a spotting guide with defined positions for the rods and an origin line. This is usually 2 cm above the lower edge of the layer, but may be 1 cm higher, especially when there is a larger amount of solvent in the chamber (higher chamber saturation).

An increase in the distance from the origin to the solvent surface (z_o) decreases the height of a theoretical plate (see Eq. (34) and the discussion in Chapter 1.1), but also decreases the effective chromatographic pathway ($z_F - z_o$) and the separation ability of the layer (Eq. (39), Chapter 1.1). In the analysis of simple two-component mixtures, one rod can even be used for two analyses. One origin is located 1 cm from the zero line marked on the rod holder and the second at a height of 5–6 cm. An example of a double analysis of a single rod is the separation of the oxidized and non oxidized methyl esters of linoleic acid[36] (Figure 13).

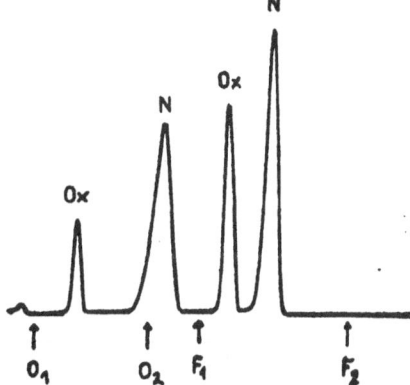

Fig. 13. Chromatogram of two samples of partially oxidized methyl ester of linoleic acid on a single Chromarod S Ox oxidized; N unoxidized methyl ester; O origins; F fronts of the partial scanning. Toluene-n-hexane system (1 : 1); recorder speed 100 mm s^{-1}, sensitivity 100 mV f. s.; 20 µg of sample applied[36]

When applying the sample, the drop formed at the tip of the micropipette needle is touched first to one side of the rod and then to the other side. This procedure is repeated until the sample has been used up.

For adjusting optimal conditions for the separation and detection in the FID, it is preferable that no concentration gradient should exist around the circumference of the rod at the origin. This requirement can be realized only by using the automatic applicator mentioned above. Point application using a micropipette does not produce controlled distribution of the sample at the starting zone and thus prechromatographic separation of the individual solutes ensues according to their polarity and the type of solvent used. Here it should be noted that Van Aerde and et al[37] were the first who applied samples to rotating Chromarods in 1979.

During manipulations with the Chromarods it is important that the fingers do not touch either the layer or the lower part of the rod holder, to avoid contamination of the rod.

The activity of the thin layers, given by the free surface energy and adsorbent surface area[19,21] is strongly affected by the relative humidity of the laboratory atmosphere. The velocity of adsorption of water on the surface of the layer is relatively fast, so that up to 50–80 % of the equilibrium adsorption can be attained in several minutes. In other words, the properties of the rods change uncontrollably during sample application, leading to deterioration of the reproducibility of the separation. Thus it has been recommended that, after sample application, the rods should be freed of adsorbed water molecules by drying for 5 min in a dessicator at 1.3 kPa (10 Torr); this is said to increase the reproducibility of the chromatography and quantitative analysis in the detector[38]. It should be noted, however, that this operation is rather tedious and still cannot ensure constant adsorbent activity, which can change rapidly during transfer of the rods into the chamber and during elution. Standardization of the thin-layer activity is a problem that can be solved either by air conditioning in the laboratory (which is quite expensive) or by adjusting the relative humidity of the atmosphere in the elution chamber. This approach will be discussed in the next section.

1.2.3.3 Elution Solvents

The choice of elution solvents is basically the same in TLC-FID as in classical TLC. It is usually the case that systems found to be useful for planar chromatography (TLC, HPTLC) can be used after a certain modification (by decreasing the contents of the more polar components) in TLC-FID. High-boiling-point solvents of a low volatility should not be used as they are difficult to remove from the rod by drying prior to the FID scanning and often cause high noise levels in the detector.

Solvents with boiling points above 100 °C, especially tetrachloroethane, pyridine, n-butanol and propionic acid, are rarely used. Toluene is an exception and is preferred to benzene because of its lower toxicity. Acetic acid is used especially for systems in which more volatile formic acid would be poorly soluble. It is usually used in amounts not exceeding 1 % and is relatively easily removed during drying in the presence of other solvents. The boiling points of the most common solvents are listed in Table 4.

Table 4 Some Physicochemical Characteristics and Elution Strengths of Solvents Useful for TLC-FID

Solvent	b. p. °C	$\gamma \times 10^3$ N m^{-1}	$\eta \times 10^3$ Pa s	$1/V_m$ $\times 10^2$	n_b	ε°	
						SiO$_2$	Al$_2$O$_3$
n-pentane	36.1	16.0	0.24	0.87	5.9	0.00	0.00
n-hexane	68.6	18.4	0.32	0.76	6.8	0.00	0.00
n-heptane	98.4	20.4	0.33	–	–	0.00	0.00
cyclohexane	81.0		1.00	0.93	6.0	0.03	0.04
carbon tetrachloride	76.7	27.0	0.97	1.04	8.6	0.14	0.18
toluene	110.8	28.1	0.59	0.94	6.8	0.22	0.29
benzene	80.1	28.9	0.65	1.13	6.0	0.25	0.35
chloroform	61.1	27.1	0.57	1.26	5.0	0.31	0.40
dichloromethane	40.7	26.5	0.44	1.57	4.1	0.31	0.42
tetrahydrofuran	65.6		0.55	1.23	5.0	0.35	0.45
diethyl ether	34.6	17.0	0.23	0.96	4.5	0.38	0.38
ethyl acetate	77.1	23.9	0.45	1.02	5.7*	0.45	0.58
acetone	56.2	23.7	0.32	1.36	4.2*	0.47	0.56
acetonitrile	81.6	29.3	0.34	1.91	10.0	0.50	0.65
1-propanol	97.2	23.8	2.09	1.34	8.0*	0.63	0.82
2-propanol	82.0	21.7	2.27	1.34	8.0*	0.63	0.82
ethanol	78.4	22.8	1.20	1.71	8.0*	0.68	0.88
methanol	64.7	22.6	0.60	2.49	8.0*	0.73	0.95
acetic acid	118.2			–	–	$\gg 1$	$\gg 1$
formic acid	100.7			–	–	$\gg 1$	$\gg 1$
water	100.0	73.1	1.00	–	–	$\gg 1$	$\gg 1$

* For silica gel and a value of $\varepsilon^\circ \geq 0.38$, $n_b = 10$, γ = surface tension (20 °C), η = viscosity (20 °C), n_b = molecule surface area (in units of 0.085 nm^2), V_m = molar volume (cm^{3}mol^{-1}).

The individual solvents are listed in eluotropic series[39], for evaluation of the magnitude of the interaction with the adsorbent surface. Solvents are ranked in Table 4 according to their elution strengths ε°, defined as the ratio of the adsorption energy of the eluent molecule E_{sa} and the surface occupied by an eluent molecule on the surface of the sorbent (A_e)[21]:

$$\varepsilon^\circ = E_{sa}/A_e \qquad (40)$$

This is an empirical factor whose values have been determined experimentally. The scale of the values of the elution strengths of the individual eluents is related to the elution strength of n-pentane, which is considered to have the lowest value ($\varepsilon^\circ_{pentane} = 0.00$). Later, however, it was found that perfluoroalkanes are even weaker eluents than n-pentane and thus their elution strengths are negative with respect to n-pentane ($\varepsilon^\circ \sim -0.25$). The higher the ε° value of the eluent, the greater the degree to which the adsorbed solute is displaced by the eluent from the surface of the thin layer and the greater its R_F value in the given chromatographic system. The eluent strength also depends to a certain degree on the type of sorbent, so that the order of the individual solvents in the eluotropic series found for silica gel is somewhat different from that for alumina.

Thin-layer chromatography is mostly carried out using multi-component elution systems, so that the selectivity of the separation is increased (see Eq. (38), Chapter 1.1). The use of mixed solvents in thin-layer elution is always problematic, as a relatively small change in the composition can lead to a large change in the selectivity and thus also in the reproducibility of the analysis. A further complicating factor is termed "demixing", i. e. a change in the composition of the mobile phase during elution as a result of preferential sorption of the more polar component(s) in the lower region of the thin layer. After depletion of the given component, a sudden change in the composition of the system occurs with formation of a secondary front.

The elution strengths of mixed solvents can be calculated only for binary mixtures assuming that the demixing effect does not occur and that interactions between the eluent and solute are negligible in both the mobile and stationary phases. Further, it is assumed that the molecules of both solvents are roughly of the same size. The following relationship can then be used for the elution strength of a binary mixture[21]:

$$\varepsilon^\circ_{ab} = \varepsilon^\circ_a + \frac{\log\,(X_b 10^{a(\varepsilon^\circ_b - \varepsilon^\circ_a)n_b} + X_a)}{an_b} \tag{41}$$

where ε_a is the elution strength of the weaker eluent, n_b is the surface area of the sorbed molecule of the stronger eluent, a is an activity parameter of the sorbent and X_a and X_b are the corresponding mole fractions:

$$X_a = \frac{\%a \cdot (1/V_a)}{\%a \cdot (1/V_a) + \%b\,(1/V_b)} \tag{42}$$

where

$$X_b = 1 - X_a \tag{42a}$$

and V_a and V_b are the molar volumes of a and b, i. e. of the two components of the binary system.

Eq. (41) indicates that the elution strength of a binary mixture depends not only on the values of the elution strengths of both components, but also on the activity of the thin layer, which is a function of the relative humidity, among other things. The values of ε°_{ab} for some binary mixtures for two activity levels are given in Table 5 and Figure 14. It can be seen that the elution strength of a mixture of two solvents increases primarily in the region of low concentrations of the more polar component and that the effect of different sorbent activities is also most marked in this region. The sensitivity of the elution strength to small changes in the concentration of the more polar component is reflected in marked variation of the R_F values in the separation of samples using insufficiently pure eluents. A typical example is chromatography using commercial chloroform,

Table 5 The Elution Strengths of Some Binary Mixtures of Elution Solvents (ε_{AB}°) for Silica Gel for $a = 0.7$ (columns 1 to 4) and $a = 0.38$ (columns 5 and 6)

Vol. % more polar component	n-hexane diethyl ether	benzene acetone	benzene methanol	chloroform ethanol	n-hexane diethyl ether	chloroform ethanol
0.0	0.00	0.25	0.25	0.31	0.00	0.31
0.5	0.08	0.26	0.45	0.39	0.02	0.33
1.0	0.12	0.27	0.49	0.42	0.03	0.34
2.0	0.14	0.29	0.54	0.46	0.05	0.37
3.0	0.16	0.30	0.56	0.48	0.08	0.39
5.0	0.21	0.32	0.59	0.53	0.11	0.42
10.0	0.25	0.36	0.63	0.55	0.16	0.47
15.0	0.26	0.38	0.65	0.58	0.19	0.51
25.0	0.30	0.40	0.68	0.61	0.24	0.56
50.0	0.34	0.43	0.71	0.65	0.31	0.62
75.0	0.36	0.45	0.72	0.67	0.34	0.66
100.0	0.38	0.47	0.73	0.68	0.38	0.68

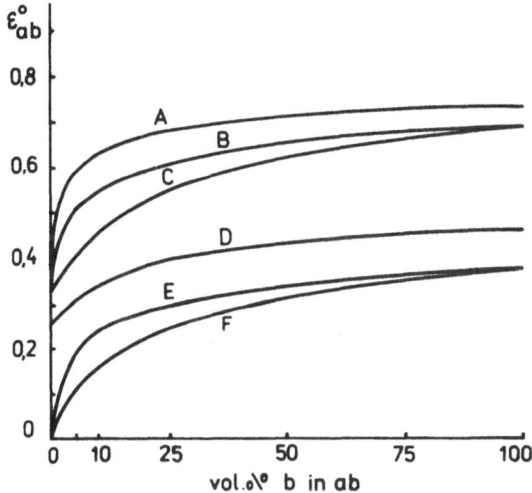

Fig. 14. Elution strength ε_{ab}° of binary mixtures, calculated for silica gel ($a = 0.7$ or 0.38). A benzene-methanol ($a = 0.7$); B chloroform-ethanol ($a = 0.7$); C chloroform-ethanol ($a = 0.38$); D benzene-acetone ($a = 0.7$); E n-hexane-diethyl ether ($a = 0.7$); F n-hexane-diethyl ether ($a = 0.38$)

which is stabilized with ethanol. The elution strengths of some batches of chloroform can lie in the range 0.31 to 0.46 (at $a = 0.7$). It is thus necessary that ethanol be removed from the chloroform (e. g. by addition of ca. 2 % phosphorus pentoxide) prior to preparation of the elution mixture. The chloroform should also be redistilled.

Solvent systems with the same elution strengths form equieluotropic series[21]. It follows from Table 5 that such series include, for example, a mixture of 75 % n-hexane and 25 % diethyl ether (v/v) and 97 % benzene and 3 % acetone. Both mixtures have the same elution strength ($\varepsilon^\circ = 0.30$). The practical importance of equieluotropic systems should not be overestimated. The selectivity of the separation also depends on secondary effects that are a function not only of the solvent composition, but also of the sorbent and solute[40]. It is very difficult to quantify these dependences and thus the experimental approach remains the best way to choose the optimal eluent.

An important property of elution solvents is the rate at which they rise in the thin layer, given by the basic flow equation (Eq. (12), Chapter 1.1). The flow of the mobile phase in the thin layer depends on the properties of the solvent, primarily the permeation factor (γ/η), as well as on the structure of the thin layer (Eq. (14), Chapter 1.1). The results obtained from measurements of the rate of rise of the most important solvents in new Chromarods (after scanning twice in the FID) are listed in Tab. 6 and compared with the flow constants found by other authors for flat silica gel layers. The time required to develop the Chromarods to a height of 5 cm and 10 cm is also given. It is surprising that the flow constant values for Chromarods are practically independent of the type of active sorbent and are much lower than the flow constants found for HR and 60_{F254} silica gels. It can be assumed that the flow-rate in this kind of thin layer is determined primarily by its frit structure. It should be added that the flow constants for the solvents determined using Chromarods in saturated chambers are comparable only with the constants valid for Merck 60_{F254} silica gel. The values for HR silica gel were measured in an unsaturated chamber and can be compared with the other values only after multiplication by a correction factor ξ_v, considering the effect of saturation (see Eqs. (10) and (14), Chapter 1.1.).

It is apparent from Eq. (14) that the flow constant is proportional to the ratio of the surface tension and viscosity (permeation factor) to the capillary radius in the thin layer and to the degree of saturation of the chamber. As both the latter quantities are usually unknown, they are included in parameter k_2 and Eq. (14) is simplified to yield

$$x = k_2 \cdot \gamma/\eta \tag{43}$$

where

$$k_2 = \frac{x}{\gamma/\eta} = k_1 \cdot r\xi_v \tag{43a}$$

ξ_v is the correction factor defined in Eq. (10), Chapter 1.1. Parameters k_2 calculated from the data in Table 6 and given in Table 7 again reflect the structural similarity of the thin layers of all three types of Chromarods and their difference, in principle, from the structure of planar silica gel layers. Differences in the parameters k_2 are relatively small for the individual solvents and vary from

Table 6 Flow Constants (x mm^2 s^{-1}) and Development Times (min) for Chromarods in Some Solvents at 20 ± 1 °C in a Saturated Iatron Chamber, $14 \times 4 \times 17$ cm[36]. For comparison, the values for planar silica gel HR layers (unsaturated chamber S)[21] and HPTLC Merck 60 F$_{254}$ (saturated chamber N)[41] are given.

Solvent	Flow constant (x)					$z_F - z_0$ Average development time (min)	
	Chromarods			Silica gel			
	S	S II	A	HR	60 F$_{254}$	5 cm	10 cm
acetone	9.3	9.3	9.9	10.9	16.2	6.5	22
diethyl ether	8.3	8.2	8.2	11.7	15.3	7	24
water	8.3	8.2	7.9	–	–	7	25
ethyl acetate	7.0	7.4	7.3	8.8	12.1	8.5	29
n-hexane	6.7	6.7	7.1	10.9	14.6	9	30
n-heptane	6.6	5.8	6.6	8.5	11.4	9	30
chloroform	6.2	6.3	5.8	7.2	11.6	10	33
toluene	6.0	6.0	5.8	8.7	11.0	10	33
benzene	5.9	6.0	6.2	8.2	10.4	10	34
methanol	3.1	3.3	3.0	6.8	7.1	19	65
ethanol	2.5	2.5	2.5	3.2	4.2	24	80

*Calculated from the relationship $\dfrac{z_i^2}{x}$ for $z_0 = 1$ cm on Chromarods S, S II and A (approximate values).

Table 7 Permeation Factors γ/η and Parameters k_2 in Eq. (43), Calculated from the Data in Table 6

Solvent	γ/η	Parameter k_2				
		Chromarods			Silica gel	
		S	S II	A	HR	60 F$_{254}$
acetone	74.1	0.13	0.13	0.13	0.15	0.20
diethyl ether	73.9	0.11	0.11	0.11	0.16	0.21
water	73.1	0.11	0.11	0.14	–	–
ethyl acetate	53.1	0.13	0.14	0.12	0.17	0.23
n-hexane	57.8	0.12	0.12	0.11	0.19	0.25
n-heptane	61.8	0.11	0.10	0.12	0.14	0.18
chloroform	47.5	0.13	0.13	0.12	0.15	0.24
toluene	47.6	0.13	0.13	0.12	0.18	0.23
benzene	44.5	0.13	0.13	0.14	0.18	0.23
methanol	37.7	0.08	0.09	0.08	0.18	0.19
ethanol	19.0	0.13	0.13	0.13	0.17	0.22
average	–	0.12	0.12	0.12	0.17	0.22

the mean value of 0.12 by $\pm$ 8 %, while the scatter of the corresponding values for silica gel HPTLC 60 F$_{254}$ is almost twice as high (Figure 15). The difference between the two types of thin layers is probably a result of the reduced ability of Chromarods to adsorb solvent molecules from the gaseous phase during elution, reflected in the lower value of correction factor ξ_v and its smaller dependence on the type of solvent. The reduced sorption capacity of Chromarods is apparently connected with the smaller volume concentration of active sorbent (silica gel, alumina) in the thin layer which, in contrast to classical layers, is "diluted" by the non-active (or less active) glass carrier.

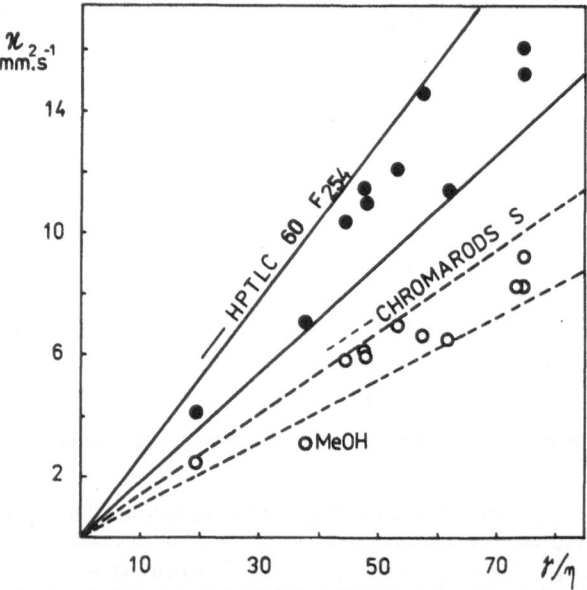

Fig. 15. Dependence of flow constant x on the permeation factor. Constructed using data in Tables 6 and 7 for Chromarods S and silica gel Merck HPTLC 60 F_{254}

Methanol occupies a rather special position in the series of solvents studied; its k_2 parameter is perceptibly lower than average (see Figure 15). This anomalous behaviour can be explained by solvation of the silica gel (and probably also the glass carrier) by methanol, reflected in a decrease in the effective capillary diameter and a decrease in the rise rate.

Ethanol and water are useful polar eluents often employed in TLC-FID. The flow-rate of these solvents in Chromarod layers almost corresponds to the theoretical value (Eq. (43)).

The parameter k_2 can be used to evaluate the quality of older Chromarods (after 50 to 100 analyses). This value is determined from the flow-rate for acetone and is compared to the original value obtained by measuring the same series of unused rods. If the k_2 value for the rods used is more than 0.02 units lower than the initial value, then the capillaries are partially blocked and the rods should be washed in dilute hydrochloric acid (see the above section on Chromarods). If this value is 0.02 units higher, then the density of the active sorbent has decreased and the separation ability of the thin layer is reduced[36].

1.2.3.4 Elution

The selected solvent system must be prepared from freshly distilled solvents, some of which (especially chloroform, see above) should be freed of foreign additives. The development tanks must be cleaned regularly. The amount of solvent depends on the chamber size used; data are given in section 1.2.2.3.

The frequency with which the solvent system is changed depends on the number of elutions and on the component volatility. Less volatile solvent systems can be used for up to a week without noticeable change in their separation abilities, i.e. for 25–30 analyses. More volatile systems (e.g. a mixture of n-hexane and diethyl ether) are usually changed daily.

Complete saturation of the elution chamber by solvent system vapours is just as important in TLC-FID as in classical paper chromatography or TLC, even though no edge effect can appear when using cylindrical thin layers.

In a chamber lined with filter paper, equilibrium saturation with low-boiling solvents (b. p. < 100 °C) is attained within five minutes, while chambers with unlined walls require two or more hours for the upper part of the chamber to become saturated[21]. In unsaturated chambers the flow-rate of the mobile phase in the thin layer decreases substantially and elution with low-boiling point solvents may even result in stopping of the solvent front and marked loss of solvent molecules from the thin layer into the chamber space. This process leads to accumulation of the more mobile components of the solute close to the front and to undesirable deformation (Figure 16).

Peak anomaly at the front can then lead to different responses for the separated zones in the FID and a decrease in the reproducibility of the separation. The responses of substances with lower R_F values are less dependent on chamber saturation, except for the least mobile component. The mobility of all the components in a saturated chamber is clearly greater (Table 8)[36].

Table 8 The Effect of Saturation of the Elution Chamber on the R_F Values, FID Responses and Reproducibility of the Determination of Tripalmitine (TG), Palmitic Acid (FA), 1,2-Dipalmitine (1,2-DG) and Monopalmitine (MG)[36]. Average of 10 determinations. Analysis conditions given in caption to Figure 16.

	TG			FA			1,3-DG			MG		
	R_F	Res-ponse	C V %	R_F	Res-ponse	C V %	R_F	Res-ponse	C V %	R_F	Res-ponse	C V %
A	0.72	1.10	12.6	0.63	0.96	8.9	0.44	0.91	3.6	0.14	1.02	3.7
B	0.79	0.74	3.8	0.67	0.97	3.9	0.56	0.99	0.9	0.19	1.18	0.9
C	0.79	0.72	3.4	0.67	0.94	2.6	0.56	0.99	1.0	0.19	1.12	1.5
D	0.75	0.80	3.6	0.65	0.96	1.6	0.50	0.99	0.6	0.18	1.08	3.1
E	0.74	1.09	3.2	0.64	0.92	8.9	0.48	0.99	1.8	0.15	1.02	3.1

A-unsaturated (latron) chamber; B-chamber lined with filter paper; C-chamber containing ten Chromarods; D-chamber containing three Chromarods; E-chamber containing one Chromarod.

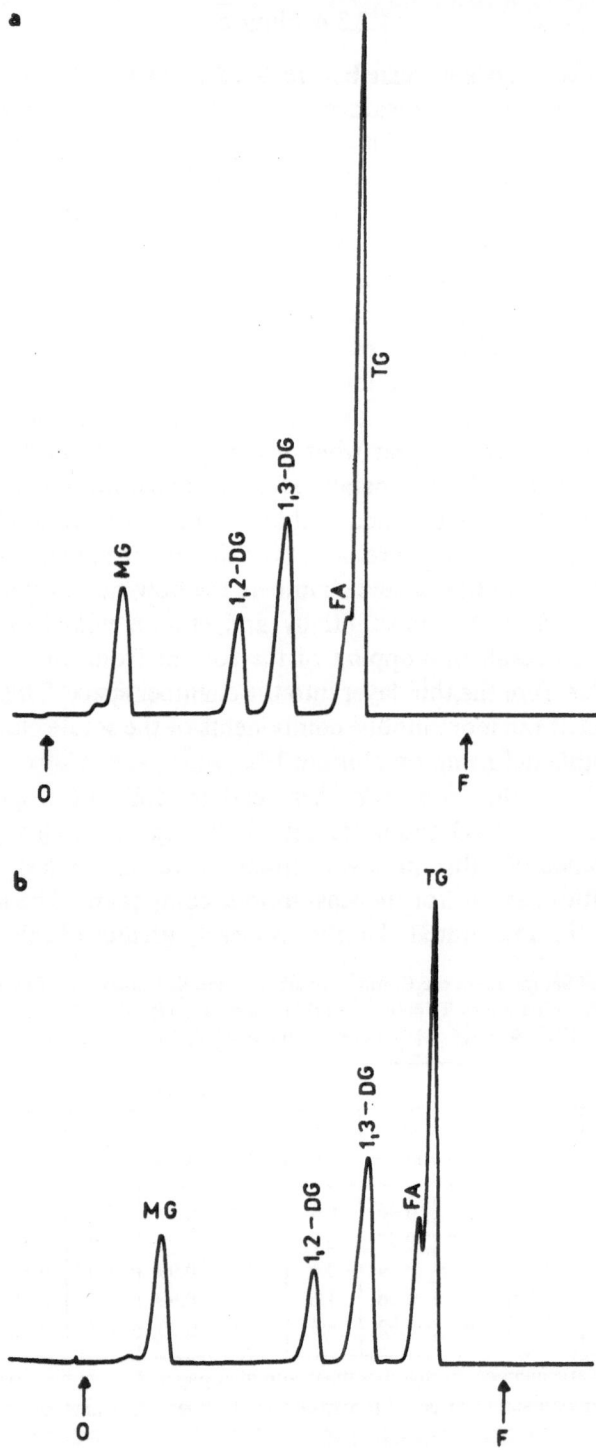

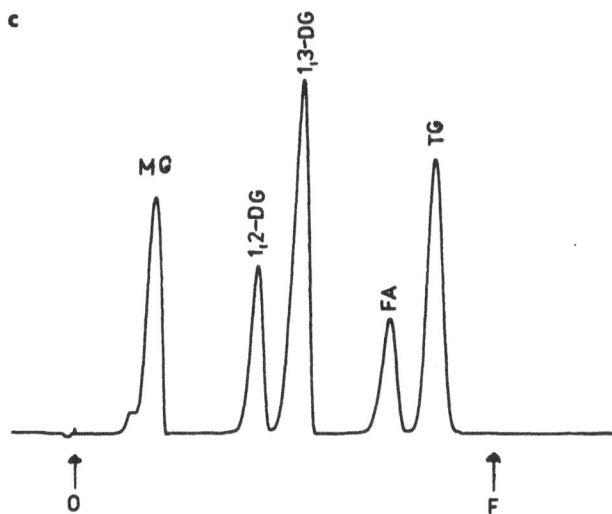

Fig. 16. Effect of chamber saturation on the peak shape and separation of components close to the front. TLC-FID of a mixture of tripalmitin (TG), dipalmitin (1, 2 and 1, 3 DG), monopalmitin (MG) and palmitic acid (FA) using a benzene-ethyl acetate-acetic acid system (95 : 3 : 2), a unsaturated chamber; b saturated chamber with two Chromarods S; c chamber saturated by lining with filter paper

Chamber saturation can be readily achieved by placing a piece of filter paper on the back wall or side walls. A set of used Chromarods can fulfill the same function. It is apparent from Table 5 that sufficient saturation could be achieved using fewer than 10 rods placed regularly in a holder located in the solvent and leaning against the back wall of the chamber. If filter paper is used (preferably chromatographic), it must be cleanly cut to prevent contamination of the solvent and atmosphere by cellulose particles. Microscopically small cellulose fibres are very easily picked up on thin layers and burn during the FID detection to form sharp spikes that considerably decrease the quality of the chromatographic recording. Prior to placing in the chromatographic chamber, the filter paper should be freed of soluble impurities by washing in the solvent system. These problems are not encountered with used Chromarods.

After placing the rods in the saturated chamber, molecules of the gas phase are gradually condensed on the thin layer and slowly change its separation ability. Condensing solvent molecules replace water molecules in the thin layer surface and thus increase its activity. If the laboratory is not air-conditioned with a constant relative humidity and temperature, then the elution procedure described above results in variation of the activities of the rods and uncontrollable changes in the composition of the mobile phase and thus in low reproducibility of the analysis. Procedures are recommended in recent papers involving preloading of the rods in a special chamber (see Figure 8) and rod preconditioning in a chamber with standard relative humidity[42,43]. This can be adjusted directly in a developing chamber such as the suspended development tank type (Figure 8) by placing a further vessel containing a saturated solution of a suitable salt in the chamber (Table 9) or containing an aqueous solution of sulphuric acid, calcium chloride or sodium hydroxide (Table 10). Adjustment of the relative humidity using an aqueous solution of sulphuric acid cannot be recommended as the vapour pressure of the acid is quite high at higher concentrations and sulphuric acid molecules can be adsorbed on the thin layer, changing its properties and chemically changing the solute composition. Conditioning and preloading of the rods prolong the analysis time by 10 to 30 min and thus decrease by one third to one half the number of analyses that can be completed in a given time interval.

Chromarods can be developed by ordinary one-step elution, by repeated elution using a single solvent and by stepwise elution with variously polar systems. The rods are dried for a short time between individual elutions in the second and third methods (1 min at 50–100 °C), depending on the volatility of the solvents.

Table 9 Relative Humidity Above Saturated Solution of the Salts at 20 °C

Solid Phase	Relative humidity %	Solid Phase	Relative humidity %
$ZnCl_2 . 1,5 H_2O$	10	NH_4Cl	79
CH_3COOK	23	KCl	85
$CaCl_2 . 6 H_2O$	33	$KHSO_4$	86
$K_2CO_3 . 2 H_2O$	44	$BaCl_2 . 2 H_2O$	88
$NaHSO_4 . H_2O$	52	$ZnSO_4 . 7 H_2O$	90
$NaBr_2 . 2 H_2O$	58	$Na_2CO_3 . 10 H_2O$	90
$NaNO_2$	66	$Na_2SO_3 . 7 H_2O$	95
$NaCl$	76	K_2SO_4	97

Table 10 Relative Humidity Above Aqueous Solution of Mixtures of Sulphuric Acid, Sodium Hydroxide and Calcium Chloride at 25 °C (mass %)[42]

Relative humidity %	H_2SO_4	NaOH	$CaCl_2$	Relative humidity %	H_2SO_4	NaOH	$CaCl_2$
100	0.00	0.00	0.00	50	43.10	28.15	35.64
95	11.02	5.54	9.33	45	45.41	29.86	37.61
90	17.91	5.83	14.95	40	47.71	31.58	39.62
85	22.88	13.32	19.03	35	50.04	33.38	41.83
80	26.79	16.10	22.25	30	52.45	35.29	44.36
75	30.14	18.60	24.95	25	55.01	37.45	–
70	33.09	20.80	27.40	20	57.76	40.00	–
65	35.80	22.80	29.64	15	60.80	43.32	–
60	38.35	24.66	31.73	10	64.45	47.97	–
55	40.75	26.42	33.71	5	69.44	–	–

Single-step elution permits maximum separation in the R_F range 0.3–0.4 (see Figure 4 and Eq. (38), Chapter 1.1.). This approach is more useful for separation of a small number of components with similar polarities. Development can be carried out using a single-component solvent (e. g. pure chloroform in the chromatography of simple acylglycerols), or binary or more complicated mixtures. The chamber can either be saturated directly with the elution system or by using a different solvent system. An example is the separation of acylglycerols using an ethanol solution in chloroform (1.1 % ethanol v) and saturation of the chamber with pure benzene.

Repeated development (usually double or triple) using a single solvent leads to better separation of neighbouring zones only when the elution strength of the system is not too great, i. e. when the retardation factors of the separated pair are relatively small, to permit favourable influence of the number of plates (the second term in Eq. (38)). At higher R_F values, the effect of the third term in Eq. (38) $(1 - R_F)$ predominates and the resolution decreased with increasing number of elutions. The dependence between the optimal number of elutions n_{el} (opt) and the retardation factor can be defined as follows[21]

$$n_{el} \ (opt) = \frac{-1}{\ln \ (1 - \bar{R}_F)} \tag{44}$$

where $\bar{R}_F$ is the mean retardation factor for the separated pair. It follows from this equation that the optimal separation for $\bar{R}_F = 0.3$ is attained after three elutions, for $\bar{R}_F = 0.4$ after two elutions; for higher R_F values each repeated elution decreases the resolution. Repeated development, often combined with stepwise development, is often used in separation using Chromarods S with a separation number (Eq. (39), Chapter 1.1) smaller than that for S II rods.

Stepwise development is commonly used for analysis of more complicated mixtures (6 to 12 components). There are two possible methods. In the first approach, the initial elution is carried out using a more polar system with a short pathlength (on one third to one half of the rod). Polar substances are separated from non-polar substances, which are accumulated at the front; chromatographic separation is then continued in the upper, vacant part of the rod using one or more solvent systems with decreasing polarity. The second procedure is the opposite of the first. The more mobile components are separated using a solvent system of low polarity over the whole length of the rod. The more polar substances are then separated using further systems in the first third to half of the rod. These approaches can be variously combined, as will be described in subsequent chapters.

Gradient elution is used less often. Here the solvent system is gradually changed after the front has attained a certain height (at distances of 2–3 cm) by addition of a more polar solvent to the chamber without interrupting the development. The analysed substance must be applied sufficiently high up the rod (say 1–2 cm above the usual start), to prevent immersion of the start after addition of the last increment. Addition of solvent and mixing it into the solvent system requires modification of the lid of the chamber (sealed openings for the stirrer and solvent inlet), or the chamber itself (an overflow siphon at the solvent surface level). No commercial equipment has been manufactured for this purpose.

The above elution procedures are also commonly used in classical TLC; however, the number of elutions is somewhat limited because of the lower mechanical resistance of the thin layer. Because of their shape, Chromarods cannot be used for two-dimensional or controlled-flow chromatography; however, this drawback is fully compensated by the possibility of using selective pyrolysis and reactivation in the FID between individual elutions.

In stepwise development using selective scanning[35], the unique ability of the Iatroscan instrument to scan and reactivate only part of the thin layer is employed. This increases the separation capacity of the thin layer several-fold and up to 12 or even more components can be separated on a single rod.

The principle of this elution method is depicted in Figure 17.

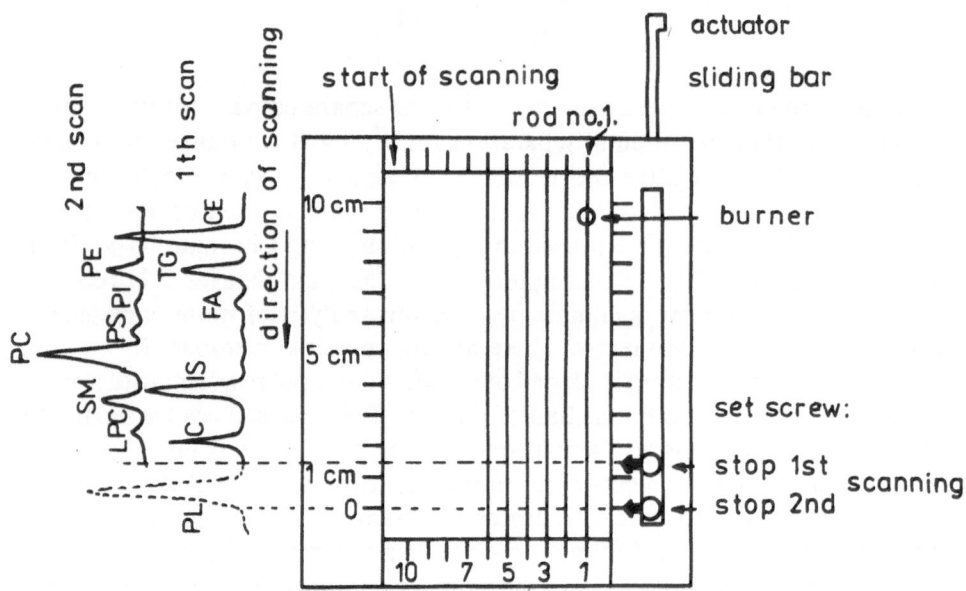

Fig. 17. Scheme of analysis of lipids from blood plasma by selective scanning. O origin; PL phospholipids; C cholesterol; IS Internal standard (palmityl alcohol); FA fatty acid; TG triacylglycerol; CE cholesterol esters; LPC lysophosphatidylcholine; SM sphingomyelin; PC phosphatidylcholine; PS phosphatidylserine; PI phosphatidylinositol; PE phosphatidylethanolamine

The rods are first developed in a solvent system of low polarity (e. g. n-hexane, diethyl ether, formic acid) and, after drying, are placed in the instrument. A screw is set to arrest travel of the Chromarod so that the FID scans only the separated components (stop position in the first detection, see Figure 17), leaving the polar components at the origin. The latter are then separated on the vacant and reactivated part of the layer using a more polar solvent system (e. g. chloroform, methanol, water). The adjusting screw is returned to the original position prior to the second scan in which the complete rod is scanned.

An alternative procedure can also be employed. After elution with a more polar solvent, the polar substances are separated on the thin layer and the mixture of non-polar substances is accumulated at the front. The rods are then placed individually (without the rod holder) in the detector, reversing the origin/front position. After scanning the separated components, the rods are developed in this reversed position using a less polar solvent system.

1.2.3.5 Drying of the Eluted Rods Prior to Detection.

Drying of Chromarods should completely remove the elution solvents so that only the separated components remain on the rod. Understandably, solvents with a greater affinity for the adsorbent must be removed from the surface at a higher temperature. For example, alcohols are hydrogen-bonded to silica gel and are incompletely removed during the drying process. Basic substances (dimethylamine, ammonium hydroxide) form stronger bonds with acid silica gel and cannot be removed from the thin layer by simple drying, and so a considerable increase in the noise level is observed during the detection. The "cleanest" chromatographic recordings, permitting quantification of even very small amounts (tenths of a microgram), are obtained by elution with systems based on n-hexane or benzene, containing larger or smaller amounts of diethyl ether or acetone. The addition of acetone is useful from this point of view especially for systems containing water.

As mentioned above, the rods can be quite easily dried in a modified glass beaker heated by an infrared lamp, and the drying can be hastened by a stream of pure nitrogen.

A commercial drying oven is also useful, as it can contain up to five rod holders (Figure 10). The temperature and drying time are dependent, to a certain degree, on the boiling points of the elution solvents, on the volatility of the separated components and on their thermal resistance. It should be recalled in this connection that substances separated on the thin layer sometimes volatilize at much lower temperatures than their boiling points. For example, glycerol perceptibly evaporates from the thin layer at only 70 °C and octadecane at 100 °C although their boiling points are much higher (290 °C and 317 °C, respectively). Thus the optimal drying conditions must be found empirically. In practice, the rods are dried for 3–20 min at a temperature of 20–170 °C. We have found, however, that the majority of solvents can be completely removed at 70–110 °C after 5–7 min.

The optimal drying time is best found experimentally by comparing the basic noise level of the clean dried rod, pre-eluted in the given solvent system, with that for the same rod prior to elution. It can be seen from the example in Figure 18 that drying for 2 min at 60 °C is insufficient even for such volatile solvents as n-hexane and diethyl ether. Five minutes at 90 °C is, however, long enough (identical noise levels for the clean layer before and after elution compare traces 4, 5 and 6 with 9, 10 and 11).

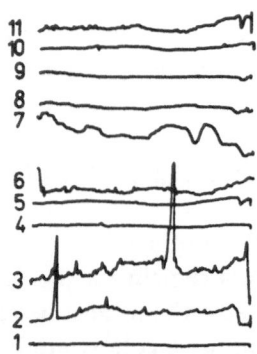

Fig. 18. The effect of drying on the background noise of Chromarods S II in the FID prior to elution (traces 1–6) and after elution for 25 min in an n-hexane-diethyl ether system (93:7). 1–3: first scanning of the clean rods, washed in distilled water and dried for 10 min at 110 °C; recorder sensitivity 100, 50 and 20 mV f.s.; 4–6: second scanning of the same rods, identical sensitivity; 7: eluted rods dried for 2 min at 60 °C, sensitivity 10 mV; 8: as 7, dried 10 min at 60 °C; 9, 10, 11: eluted rods, dried 5 min at 90 °C, sensitivity 100, 50 and 20 mV. Detector parameters: H_2 180 ml min^{-1}, air 2.1 l min^{-1}, scanning speed 0.42 cm s^{-1}

After drying, the rods are placed in a dessicator and transferred to the detector.

1.2.3.6 Flame Ionization Detection

Flame ionization detection (FID) is based on measurement of a change in the ionization current as a result of a sudden increase in the ion density flowing between the cathode (usually the burner) and anode (usually the collector electrode) after introduction of the analysed component into the hydrogen--oxygen flame[44]. The background ionization current of pure hydrogen is very low but can increase in the presence of traces of organic impurities (originating for example from the stationary phases in GLC and TLC) to a value of up to 10^{-10} A [45].

The burning flame is a system in which a very rapid chain reaction occurs to yield a large number of free radicals and excited particles. During interaction of these particles, a large amount of heat is evolved, leading to ion formation (chemionization) before the temperature drops to that of the surroundings (thermalization). Chemionization occurs primarily in the reaction zone formed by the narrow region of the flame (only a few tenths of a millimetre) at the point where intimate mixing of H_2 and O_2 occurs, i. e. in the layer rich in oxygen atoms. Reaction of neutral H, O and OH species in the hydrogen flame leads primarily to the formation of H_3O^+ ions, free electrons and OH^- anions. In addition to H_3O^+, burning of hydrocarbons results in formation of CHO^+, CH_3O^+, $C_3H_3^+$, etc. and negative ions along with electrons and OH^-, primarily C_2^-, O^-, C_3^-, CO_2^-, etc.[46]. The combustion of organic substances leads to such an increase in the flame temperature that some species exceed their ionization threshold and are ionized. This thermionization yields, for example, NO^+ ions from the combustion of nitrogen in an air stream and the alkali metal cations[46].

Chemionization and thermionization are primary ionization processes. Further ions can be formed by charge transfer, i. e. by a process in which charged particles react with neutral particles to form other charged particles, while the overall charge remains unaffected. Thus charge transfer does not lead to an increase in the ionization current. An example of charge transfer is the reaction of CHO^+ with a water molecule to form a hydrated proton (CH_3O^+) and CO.

The ionization current is decreased by recombination, involving reaction of a positive ion with an electron (i. e. direct recombination) or reaction of a cation with an anion, formed by prior reaction of a neutral species with an electron (i. e. indirect recombination). The most important type of reaction between oppositely charged particles is dissociative recombination of a hydrated proton with OH^- to form a neutral species that is very unstable and decomposes to form H_2O, OH and H.

The ion concentration (called the ion profile), measured from the burner towards the collecting electrode depends primarily on the type of ionization. The thermionic profiles have a maximum beyond the reaction zone at the point of maximum flame temperature. They then slowly decrease with decreasing temperature. On the other hand, the highest chemion concentration lies in the reaction zone and rapidly decreases outside this zone. The profiles for various thermions are very similar as they are determined primarily by the temperature gradient. The profiles of chemions vary considerably because their concentrations are given by the equilibrium between competing and subsequent reactions[46].

The flame ionization detector is sensitive to almost all organic substances. It is therefore a non-selective, universal detector. The term ionization efficiency has been introduced to characterize the degree of ionization of various substances. This is the charge produced in the flame by one gram atom of carbon. The molar response is a related quantity, defined as the charge transferred in the flame after pyrolysis of one mole of the given substance. The molar response of organic substances is about 10 coulombs per mole, so that the detection of only 10^{-6} moles of the substance appears as a change in the background ionization current by 10^{-7} A [45].

The ionizability of substances changes with the detection regime and thus the relative molar response is often used. This is the ratio of molar response values for the given substance and a reference compound, usually the paraffinic group —CH_2— (effective carbon)[46]. In chromatographic practice, mass responses are often used. The relative mass response is equal to the ratio of the magnitudes of the analogue FID signals for the same amounts of the analysed and reference substances. These signals are measured as the areas of the corresponding chromatographic peaks.

Much of the information about the effect of operational parameters on the magnitude of the ionization current has been obtained primarily in connection with gas chromatography. This data is basically also applicable to TLC-FID but it cannot be assumed that the response of a certain substance and the sensitivities of detection will be identical in GLC-FID and TLC-FID. While the whole separated fraction enters the flame in gas chromatography in the gaseous state, in TLC-FID the substance enters the flame together with the thin layer which affects the geometry of the flame and its temperature in a manner that is dependent on the velocity of its motion. The thin layer is heated during the detection, increasing the magnitude of the background ionization current and thus decreasing the sensitivity of the detection.

In the next three sections we consider a number of important parameters affecting the magnitude and reproducibility of the signals yielded by organic substances in detection on Chromarods in the FID.
These are primarily:

 – the hydrogen and air (or oxygen) flow-rates through the detector and their purities
 – the detector geometry and the rate of motion of the thin layer in the detector
 – the composition of the analysed sample.

Hydrogen and Air Flow-Rates

In contrast to GLC, the Iatroscan FID employs only two gases, hydrogen and air.

So far, surprisingly few papers have dealt with optimization of the gas flow-rates, compared to GLC-FID. The manufacturers recommend flow-rates of 160 ml min^{-1} for hydrogen and 2 l min^{-1} for air, in agreement with the results of Japanese authors[47], who found that the detector response and reproducibility are directly proportional to the hydrogen flow-rate up to 180 ml min^{-1} (Figure 19).

An increase in the flow-rate of hydrogen above 160 ml min^{-1} leads to an increase in the detector flame temperature and increases the reproducibility to a certain degree. The upper limit to the hydrogen flow-rate depends on the

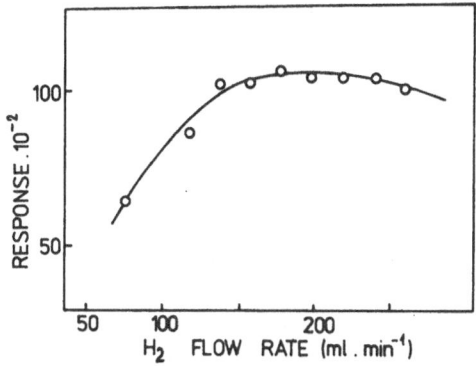

Fig. 19. Effect of the hydrogen flow-rate on the detector response. Total response for a mixture of cholesterol (0.39 μg), palmitic acid (0.43 μg), cholesterol palmitate (0.38 μg) and tripalmitin (0.37 μg). Solvent system n-hexane, diethyl ether, formic acid (54:6:0.05)[47]

scanning rate. With slow movement of the rod in the FID, partial sintering of the glass particles in the frit could occur. Bradley[48] recommends a maximum flow-rate of 180 ml min^{-1}. We have tested rates of up to 250 ml min^{-1} at a scanning velocity of 0.42 cm s^{-1} without observing any thermal destruction of the thin layer or deterioration of its separation ability[36]. An increase in the flow-rate of hydrogen (with simultaneous decrease in the flow-rate of air or oxygen) can be especially important in the TLC-FID of substances with lower carbon contents and higher halogen contents because, as follows from some papers on GLC-FID, at a certain excess of hydrogen in the detector the response to carbon decreases and the response to hydrogen halides increases[49,50]. Some closed systems with a hydrogen atmosphere, called HAFID systems (hydrogen atmosphere FID) are quite sensitive for the detection of organometallic compounds[51].

While the present design of the Iatroscan TH-10 instrument detector does not permit adjustment of the conditions described above, the sensitivity of the detector for some ions can nevertheless be increased to a certain extent by decreasing the air flow-rate, i.e. increasing the $H_2:O_2$ flow-rate ratio (Table 11)[52].

The amount of air passing through the detector cannot be precisely measured, in contrast to the hydrogen flow-rate. It is, however, only a fraction of the overall amount fed into the instrument. The largest volume of air passes between the

Table 11 Increase in the Response for Compounds Containing Chlorine by Adjusting the Hydrogen – Air Ratio. Mean mass response (average of seven determinations) and corresponding standard deviation (S D). Chromarods S, chloroform – n-hexane – acetone system (80:19:1, v/v), Scanning speed 0.42 cm s^{-1} [52].

Flow-rate 1 min^{-1}		Mean mass response ± S D			
air	H$_2$	TG[a]	PCC[b]	iso-PCA[c]	PCA[d]
2.1	0.18	1.20 ± 0.04	0.68 ± 0.03	1.00 ± 0.02	1.00 ± 0.02
1.5	0.18	1.18 ± 0.06	0.73 ± 0.04	1.00 ± 0.03	0.98 ± 0.03
1.5	0.25	1.08 ± 0.05	0.77 ± 0.05	1.02 ± 0.02	1.00 ± 0.05

a triolein; b 1-phenyl-4,5-dichloropyridazone(6); c 1-phenyl-4-amino-5-chloropyridazone(6);
d 1-phenyl-4-chloro-5-aminopyridazone(6).

inner and outer mantle of the burner body and forms a protective layer preventing entry of dust particles and other contaminants from the atmosphere into the hydrogen flame.

Under standard conditions, i. e. at a hydrogen flow-rate of 160–180 ml min^{-1} and of air of 2 l min^{-1}, a quite large diffuse flame is formed that completely engulfs the Chromarod. In the scanning of zones with higher concentrations of carbon-containing compounds at slow Chromarod velocities, the flame is sometimes orange-coloured, probably as a result of the effect of radiation from incompletely burned carbon particles. In other words, there is insufficient oxygen in the reaction zone. An increase in the oxygen flow-rate does not substantially change the ionization because the design of the burner is such that part of the added air passes between the inner and outer mantles of the burner and escapes through a vent in the detector lid. A bluish flame colour, typical for a correctly functioning FID, can be attained by decreasing the hydrogen flow-rate to 120–140 ml min^{-1} which, however, causes a decrease in the ionization current and the detection sensitivity[53].

An air consumption of about 2 l min^{-1} is quite high and thus the manufacturer provides the instrument with an air pump. However, under normal laboratory conditions, the air is contaminated to such an extent that, even after filtration through a built-in filter containing active carbon, it considerably increases the detector noise and worsens the quality of the chromatographic recording. Thus, in present practice, only compressed air from a cylinder is used. When the instrument is run constantly, up to 1000 l of air daily can be consumed. The hydrogen consumption, about 100 l daily, is more acceptable.

The purity of both gases is a basic requirement for low detector noise. Hydrogen from pressure cylinders is sufficiently pure; this may not be true of air, the quality of which is not standardized. It is thus preferable that a molecular sieve bed (5A or 13X) be placed between the pressure cylinder and the FID (in addition to the active carbon filter mentioned above).

A higher noise level of the detector can also result from other factors, including excessive solvent concentrations, smoke or dust in the laboratory atmosphere, solvent impurity and incomplete drying of the rods prior to detection, dust in the detector compartment, imperfect instrument grounding and also contamination of the burner and the collector electrode.

A common source of noise can be attributed to particles of active carbon passing through the air tubing from the filter into the burner jet. The flame-ionization detector should be cleaned regularly, at least once a month when in use. Washing is best carried out using a mixture of toluene, n-hexane and methanol (2:1:1, v/v). Acetone and chloroform should not be used as they corrode the electrode surface. Detailed instructions for cleaning can be found in the instrument manual.

Detector Geometry and Scanning Rate

The manufacturer so far provides only one type of detector, consisting of a stationary collector electrode (anode) and a burner (cathode), whose distance from the Chromarod surface can be adjusted by a screw. The collector is cylindrically shaped and has a diameter of 1.5 cm. It is placed ca 0.6 cm from the burner tip; this rather large distance has a detrimental effect on the uniformity of the electric field[53]. It is expected that a new instrument will have a different detector design to permit simultaneous sample detection in the FID and an FTID (flame thermionic detector). Preliminary tests have demonstrated that detector modification with the burner closer to the collector electrode can yield a much higher ionization current at lower hydrogen flow-rates. A decrease in the distance between the electrodes leads to an improvement in the electric field uniformity and decreases the number of recombination reactions[46].

It is known from experience with gas chromatography that the ionization rate is greater on the side of the flame towards the anode[46]. This phenomenon has also been observed during detection on Chromarods; the response of substances adsorbed on the side of the rod turned away from the flame, i.e. closer to the collector, is about 10 % higher[36].

In the Iatroscan, the FID is an open system so that, as with analogous detectors used for GLC, unfavourable effects can occur as a result of changes in the surrounding atmosphere. In addition to solvent vapours, which can considerably increase the background signal, a primary effect is excercised by fluctuations in the atmospheric pressure. The dependence of the ionization ability on common pressure fluctuations in the range 1–2 kPa can decrease the reproducibility of the analysis by up to 5 %[46]. The effect of small changes in the laboratory temperature on the ionization ability of substances in the diffuse flame is negligible[46].

One of the very important parameters in the Iatroscan detection system is the variability of the velocity of the rod movement through the flame. As follows from Table 1, the rod velocity can be varied stepwise from 0.13 to 0.73 cm s^{-1} for Mark II and from 0.25 to 0.51 cm s^{-1} for Mark III. It is apparent that, the slower the motion through the FID, the more the rod will be heated.

Sedláček et al[36] measured the dependence of the rod temperature on its velocity in the flame and distance from the burner tip. Table 12 and Figure 20 give some of the results of this study.

Table 12 Dependence of the Flame Temperature (°C) on the Rod Velocity through the Detector and on the Distance of the Rod from the Burner Surface[36]. Measured using a thermocouple at a flow-rate of 180 ml H$_2$ min^{-1} and 2.100 ml air min^{-1}.

Distance of the rod from the burner (mm)	Rod velocity (cm s^{-1})						
	0.0	0.13	0.21	0.31	0.42	0.58	0.73
0.25	605	545	490	420	360	295	260
0.40	620	560	500	420	365	305	265
0.50	645	585	515	440	375	310	275
0.80	650	590	525	450	380	315	280
1.00	655	585	520	445	380	315	275
1.90	620	530	480	425	340	280	265
3.80	505	435	395	330	290	250	225

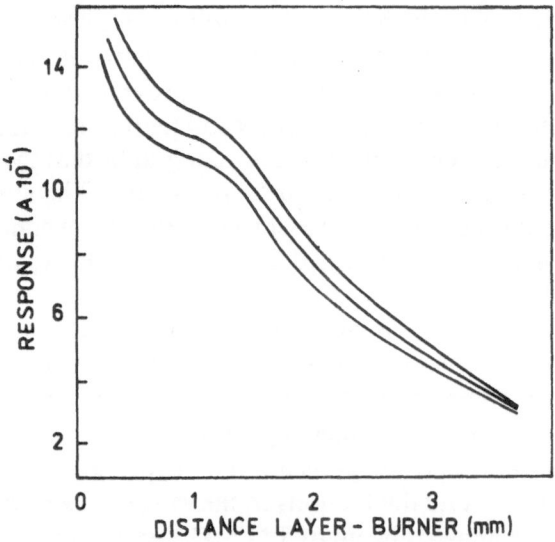

Fig. 20. Dependence of the detector response (expressed as the integrated area) on the distance of the Chromarod surface from the burner surface at a scanning speed of 0.42 cm s^{-1}. Applied: 4.8 µg of 1-monopalmitin (upper curve), 4.5 µg dipalmitin (lower curve) and 4.9 µg tripalmitin (middle curve). Developed with pure chloroform, tank saturated with pure chloroform vapours[36]

The values in the table confirm the expectation that the rod temperature will decrease with increasing velocity of the rod in the detector and increase up to distances of 0.8 to 1 mm from the burner tip. The "universal parameters" for detection, recommended by the manufacturer, i. e. a velocity of 0.3–0.4 cm s^{-1} and a distance of 0.3–0.5 mm correspond to rod temperatures of 420–440 °C at the lower rod velocity and 360–375 °C at the higher velocity. The sample response in the FID is, however, not proportional to the flame temperature; at constant scanning rate, however, it decreases with increasing distance from the burner tip (Figure 20)[36].

The decrease in the response is apparently connected with the fact that, as the Chromarods are moved further away from the burner, they leave the flame reaction zone. An increase in the scanning velocity is reflected in a higher response for the substances and higher reproducibility of the determination (Figure 21)[36].

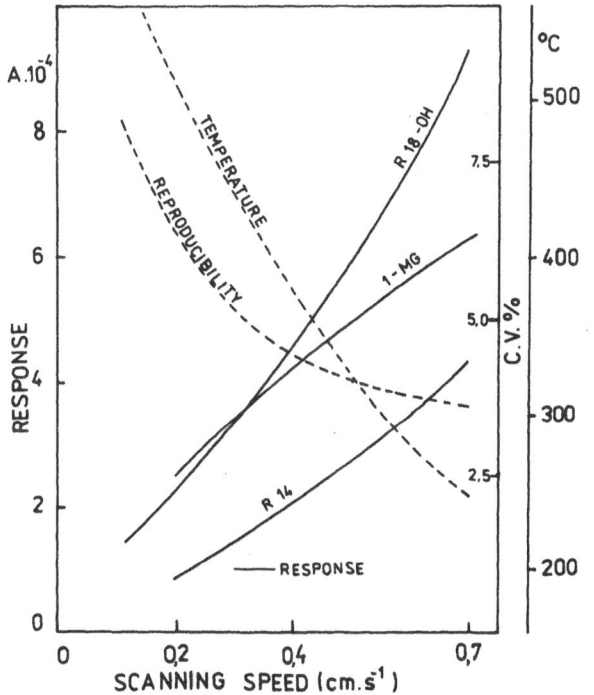

Fig. 21. Dependence of the response, reproducibility (C V %) and temperature in the pyrolysis zone on the scanning speed. 1-MG = 1-monoacylglycerol; R 14 = tetradecane; R 18-OH = octadecanol

The increase in the ionization current at higher velocities seems paradoxical; it it can be explained as follows:

– the ionization current is proportional to the velocity at which the substance enters the flame[46], or

– at higher velocities of the rod into the flame, the volatilization of the substance outside the detection zone is less intense.

On the other hand, slower movement of the rod in the flame leads to heating of a broader zone of the rod and a greater fraction of the analysed mixture can volatilize than at slower rates[36].

Crane et al in considering the reproducibility in TLC-FID, have suggested that the increase or decrease in the response at various scanning rates is a result primarily of the complexity of the kinetics of solute ionization in the FID, rather than evaporation of part of the separated zone outside the detector flame[54]. It is apparent that the degree of evaporation outside the detector depends primarily on the flame temperature and rod conductivity, in addition to the volatility of the substances. Although the conductivity of the rod has not been measured, it can be estimated from the conductivity of glass that, at a flame temperature of about 500 °C (see Table 12), the rod in the immediate vicinity of the flame is heated to 200–250 °C. At this temperature and a scanning rate of 0.42 cm s^{-1}, about 30–50 % of the substance volatilizes outside the flame zone. This fact has been verified experimentally by analysis of a mixture of mono and tripalmitates (5 µg of each component). After development of a set of 10 rods, half were passed over the flame at the given rate at a distance of 9 mm from the burner. Under these conditions, the surface was heated to a temperature of 240–250 °C without apparent ionization of the separated components. The complete set of rods was then scanned in the FID and the areas of the zones on the thermally exposed and non-exposed rods were compared. In the detection of tripalmitate, the detected difference in the areas was 30 %. This difference increased to as much as 50 % for the more volatile monopalmitate[36]. Indirect evidence for this phenomenon has been obtained by exposing Chromarods to iodine vapour before scanning in the Iatroscan. Aliphatic unsaturated components exposed to iodine gave greater responses than those unexposed. It is likely that iodine adds across a proportion of the unsaturated bonds and the derivatives formed have higher boiling points than the original compounds. This process would retard the loss through premature vaporization of the more volatile material from the Chromarod prior to entering the flame[55].

The effect of the scanning rate on the FID parameters also appears in the analysis of mixtures of various types of substances, such as aliphatic esters and heterocyclic compounds (Tables 13 and 14).

It can be seen here that the best reproducibility was attained at the highest scanning rate with, however, incomplete combustion of the less volatile component (triolein in Table 13)[52]. This fact should be considered in parameter optimization. Complete pyrolysis of triacylglycerol can be attained either by increasing the hydrogen flow-rate or decreasing the amount of substance applied[36]. The absolute mass response, expressed in terms of the area of the FID signal corresponding to 1 µg of substance, increases with the scanning rate for

Table 13 Response and Reproducibility of the Determination of Triolein and Pyrazone Derivatives at Various Scanning Speeds. Flow-rate: H_2 180 ml min^{-1}; air 2.1 l min^{-1}. Elution system and compositions of PCC, PCA and i-PCA are given in Table 11[52].

Scanning speed cm s^{-1}	Triolein		PCC		PCA		iso-PCA	
	$1/K_F$	SD	$1/K_F$	SD	$1/K_F$	SD	$1/K_F$	SD
0.21	1.25	0.11	0.58	0.03	1.08	0.03	0.99	0.03
0.31	1.35	0.13	0.60	0.03	1.02	0.04	0.99	0.03
0.42	1.20	0.04	0.68	0.03	1.00	0.02	1.00	0.02
0.42*	1.20	0.04	0.68	0.03	1.00	0.02	1.00	0.02
0.73	1.17	0.02	0.81	0.01	1.00	0.01	0.99	0.01
0.73*	1.20	0.02	0.81	0.01	0.98	0.01	0.99	0.01

$1/K_F$ mean mass response (K_F is a correction factor calculated from the corresponding response areas and the contents of the given components in the model mixture); SD standard deviation of the individual determinations calculated from the results of five analyses.
* calculated from the sum of the areas for two subsequent detections on the same rod.

Table 14 Responses (A and $1/K_F$) and Reproducibility (SD) for the Determination of Ethyl Stearate, 4-Methylmorpholine-hydrochloride (RW_1), and 4,4-Dimethylmorpholinechloride (RW_3). Acetone — dietyl ether — methanol system (62:30:8 v/v). Conditions as in Table 13[56]. A relative area (from the integrator) for 1 μg of analysed component.

Scanning speed	Ethyl stearate			RW_1			RW_3		
cm s^{-1}	A	$1/K_F$	SD	A	$1/K_F$	SD	A	$1/K_F$	SD
0.21	14.0	1.46	0.02	3.3	0.34	0.04	6.9	0.72	0.03
0.31	18.5	1.43	0.04	4,8	0.37	0.03	10.4	0.80	0.05
0.42	19.5	1.40	0.04	6,2	0.45	0.04	11.0	0.79	0.06
0.73	21.1	1.40	0.03	9.0	0.60	0.02	11.7	0.78	0.02

all components (Table 14), while the relative mass response either does not change with changing velocity of the rod motion in the FID or even decreases (triacylglycerol, ethyl stearate), or increases (PCC in Table 7, RW_1 in Table 14). This phenomenon is again related to the composition of the analysed substances; more volatile or thermolabile compounds have a lower relative response, which is clearly dependent on the temperature of the rod in the vicinity of the detector.

The Effect of the Composition of the Analysed Sample

The response to analysed substances in TLC-FID, similar to that in GLC-FID, is strongly affected by their compositions. Hydrocarbons have the greatest ionization abilities, while compounds containing oxygen, sulphur, phosphorus or halogens yield lower responses, proportional to the content of effective carbons (see above).

Attempts to quantitatively compare and tabulate the contributions of the individual functional groups to the overall response of the given compounds have so far been unsuccessful, as the response in TLC-FID is strongly dependent not only on the detection regime, but also on the Chromarod quality [57] and volatility of the detected sample.

Some volatile hydrocarbons yield a smaller response than less volatile hydrocarbons or compounds with a lower number of effective carbons, especially at low scanning velocities.

1.3.2.7 Methods for Quantitative Evaluation of the Results of TLC-FID Analyses

The recording of TLC-FID results is rather similar to that used in GLC-FID. Here also, the analogue signal is amplified and fed into a two-channel recorder depicting analogue and integral chromatographic signals. The analogue signal is also commonly integrated in an electronic integrator. Most integrating instruments treat data from the detection on the basis of the first derivative of the curve of the separated zones, usually termed peaks. If the value of the first derivative exceeds a selected sensitivity level, integration of the peak areas begins (Figure 22). If the first derivative changes from a positive value, through zero, to a negative value, the retention time of the peak is recorded. As soon as the negative value decreases to the sensitivity level, peak integration is ended and the result is stored in the computer memory. Integration yields the peak area, which

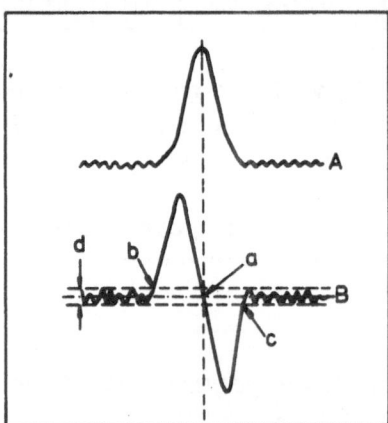

Fig. 22. Peak indication in the integrator by single differentiation of the detector signal: A analogue signal; B first derivative; (a) retention time; (b) beginning, (c) end of the peak detection; (d) selected sensitivity level (slope sensitivity)

is a quantity that is proportional to the time integral of the detector response to the analysed substance. As the integration is carried out with respect to time, the area thus obtained is not proportional to the chart rate in the recorder or plotter. After completion of the integration of the whole chromatographic curve, the results are handled in the computer or integrator to yield a report that is printed out (see also section 1.2.2).

Basic problems encountered in statistical evaluation of results have been surveyed in the literature[58] and detailed instructions are given in the manuals provided with digital integrators. These procedures are mostly designed for evaluation of the results obtained from gas and liquid chromatography, but can be modified for use in TLC-FID. It should be noted, however, that the Iatroscan instrument measures up to 10 rods in a single series, where each of these represents an independent chromatographic system. This is a basic difference in the evaluation of conclusions following the statistical evaluation of results obtained from GLC (or HPLC) and from TLC-FID, e. g. by the external standard method. In GLC (HPLC), the calibration analysis and the actual analysis are mostly carried out using a single column; in TLC-FID, different rods normally are used, although their properties are very similar. It is, in principle, possible to draw calibration curves on the basis of results obtained by analysis of standard mixtures on all 10 rods in the set, but this is too tedious and the validity of the relationships obtained in this way is limited as the separation ability of the rods and the response to components of the standard mixture vary to a certain degree with the number of analyses. Thus average values are taken to be valid for the whole set of Chromarods.

Several procedures can be used for quantitative evaluation of chromatograms.

The internal normalization method is the simplest approach, based on percentage expression of the composition of the mixture in terms of the measured areas of all the peaks on the chromatogram (A_i). Thus no data is considered in terms of the absolute amount of sample applied. The internal normalization method has two versions, employing or not employing correction factors (K_F) for the response of the substances in the detector. The version without a correction factor is used in analyses of mixtures for which the responses to the individual components in the FID are roughly identical. It is especially useful for series analysis of mixtures where the chemical compositions of all the components are known and all the components are detected in the FID. The percentage of component i in the mixture, P_i, is then obtained by division of the integrated area of peak A_i by the sum of the areas of all the peaks (ΣA_i) and multiplication by 100:

$$P_i = \frac{A_i}{\Sigma A_i} \cdot 100 \qquad (45)$$

The version employing a calibration factor has more general applicability and differs from the former method in that it considers the response of the individual components in the detector. The reciprocals of the response values, i. e. correction factors K_F, are found by the analysis of a standard mixture of a known composition on identical thin layers as those used for the analysed sample. If the response is constant over the whole range of measured concentrations, then the following relationship holds for the correction factors:

$$K_{F_i} = \frac{P_i \Sigma A_i}{100 A_i} \tag{46}$$

where P_i is the percentage content of component i in the calibration mixture, A_i is the relative area of the signal of component i in the detector and ΣA_i is the sum of the areas of the signals of all the components in the mixture. The contents of the individual components in the analyzed sample are then found using the equation

$$P_i = \frac{K_{F_i} A_i}{\sum\limits_{i=1}^{n} (K_{F_i} A_i)} \cdot 100 \tag{47}$$

where the expression in the denominator is equal to the sum of all the corrected areas in the chromatogram.

If it can be guaranteed that the dependences of the areas on the applied amounts of all the components of the calibration mixture are straight lines passing through the origin, then one of the components can be selected as a reference substance (ref) and the relative correction factors for the other components can be found:

$$K_{F_i}(\text{ref}) = \frac{c_i A_{\text{ref}}}{A_i c_{\text{ref}}} \tag{48}$$

for $K_F(\text{ref}) = 1.00$; c_i and c_{ref} are concentrations of the analysed and reference substances, respectively. Eq. (47) is then used to calculate the contents of the analysed components. This type of calculation is a common part of the programs built into electronic integrators. If the chromatographic parameters of all 10 rods in the set are roughly identical, then a standard mixture is applied to the first four rods. This mixture contains the same components as the analysed sample and the ratios of the individual components in the standard mixture should be roughly the same as in the sample. After analysis of the first rod the detection is stopped and the individual concentrations of the components are assigned to the individual retention times in the integrator memory. The integrator is then programmed so that the subsequent three analyses are carried out in the calibration mode and the next six in the analysis mode. After switching on the detector, the integrator then calculates correction factors for the response from the results of the first three analyses. These factors are then assigned to the peaks in the sample chromatograms according to their retention times. The analysed signal for the detection of each sample is treated independently and the print-out gives the areas of all the peaks, their retention times and the uncorrected and corrected percent contents of the components of the analysed samples and their correction factors.

Internal normalization is simple and is not dependent on either the volume or the concentration of the applied sample. However, it can be used only for analysis of samples where all the components are known and detectable in the FID.

The absolute calibration method is also called the external standard method and is based on application of known weights of analysed sample and standard, chromatographing and detection under identical conditions and comparison of the sizes of the corresponding peaks. The standard can be any substance that is chromatographed in the given solvent system and whose response in the FID in the selected range (0.5–20 µg) is not dependent on the amount applied. It is useful to employ one of the components of the analysed mixture, a substance with a medium retardation factor. The external standard method is used mostly in analysis of samples where a large number of components prevents inclusion of an internal standard. The external standard may be a standard mixture containing the same components as the analysed sample. Then the determination procedure is similar to that in the internal normalization method, the only difference being that it is necessary to know both the amount of components of the standard mixture applied as well as the overall amount of sample applied. There are two possible evaluation procedures, the method of direct comparison of the areas of the peaks of the analysed components with the area of the standard and the calibration curve method.

The first variation is based on the assumption that there is a constant response for all the analysed components in the given range of applied amounts. When the correction factors are known (determined basically in the same way as in the internal normalization method), the amount of component in the sample can be found from the relationship

$$m_i = \frac{K_{F_i} A_i m_{st}}{K_{F_{st}} A_{st}} \tag{49}$$

where the subscript $_{st}$ refers to the standard. If the analysed component and the standard are identical, Eq. (49) can be simplified to give

$$m_i = \frac{A_i m_{st}}{A_{st}} \tag{50}$$

When using an electronic integrator, the procedure is similar to that employed in the internal normalization method. After selection of the external standard calibration method, the standard sample (or mixture) is first analysed and the weights of the individual components in the standard mixture in the applied amount (usually 1 μl) are assigned to the retention times. The integrator calculates correction factors and multiplies the areas of the peaks obtained during the analysis by these factors. In this method, the sample composition is not recalculated to mass per cent, but is given in terms of the mass of all or only some of the components in the sample.

In the second approach, a graph is drawn to show the dependence between the applied amount of a standard substance (or substances) and the areas of the corresponding peak(s) in the chromatogram. The calibration curve should be linear and should pass through the origin. Here it is again useful if the components of the standard and analysed mixture are identical; the amount of substance applied can then be read from the calibration curve after determination of the area A_i.

An important factor affecting the precision of the construction of calibration curves in TLC-FID is the reproducibility of the application of a constant volume of calibration solution. Even with the greatest care, the actual amount of sample applied varies by at least $\pm 5\%$, which understandably has a detrimental affect on the magnitude of the standard deviation of the determined dependence of the area of the FID signal on the concentration.

Consequently, the external standard method is used only exceptionally in TLC-FID.

The internal standard method is based on addition of a known amount of a standard substance to an analysed solution, where the dependence of the response on the concentration of the standard substance is known and linear. In addition to these basic conditions, the internal standard should fulfill the following requirements:

– it should be readily soluble in the same solvents as the analysed sample

– it should be well separated from the other components in the elution system used for analysis of the given samples

– it should not react chemically with any of the components of the analysed material or affect their physical properties (e. g. through complex formation, etc.)

– it should have a similar response to those for the sample components.

The amount of standard added should be selected so that the size of the standard peak on the chromatogram is comparable with the area of the peak of the component present in the greatest amount and, simultaneously, a concentration that would decrease the resolution of the separated zones should not be exceeded. In other words, about 3–10 μg of internal standard should be applied.

This method is advantageous in that the reproducibility and accuracy of the determination do not depend on the amount of sample applied to the rod; the precision of the application plays a lesser role than in the external standard method.

The internal standard method has two variants. In the direct comparison method, a volume of standard solution equal to V_{st}, with concentration c_{st}, is added to a known volume of analysed sample V_i with unknown concentration c_i. When the correction factors K_{Fi}, K_{Fst} are known, the concentrations of the analysed components are calculated from the relationship

$$c_i = \frac{K_{Fi}}{K_{Fst}} \cdot \frac{A_i}{A_{st}} \cdot \frac{V_{st}}{V_i} \cdot c_{st} \tag{51}$$

The internal standard can also be added as a known weight to the dry sample; with a sample weight of m_{as} and internal standard weight of m_{st}, the mass percent of component i in the sample is then

$$m_i(\%) = \frac{K_{Fi}}{K_{Fst}} \cdot \frac{A_i}{A_{st}} \cdot \frac{m_{st}}{m_{as}} \cdot 100 \tag{52}$$

regardless of the sample dilution by solvent prior to application to the layer. Most common types of integrators have a program for evaluation of the variant with direct comparison where the K_F value for the components is independent of the amount of sample with the internal standard that is applied. Correction factors are usually related to the value of the correction factor for the internal standard ($K_{Fst} = 1.00$) and are found by analysis of a sample with known contents of the analysed components and internal standard.

In the calibration curve methods, several standard mixtures are prepared containing roughly the same amount of internal standard and the other components in various mass ratios. The standards are prepared either by weighing the pure components or by mixing solutions of known concentrations. After analysis, the results obtained are converted by linear regression to the dependence of the ratio of the areas of the analysed components (A_i) and internal standard (A_{st}) to their mass ratio (M_i/M_{st}) in the calibration mixtures. The calibration curve obtained has the general form

$$A_i/A_{st} = a + b \, M_i/M_{st} \tag{53}$$

where b is the slope of the straight line and a is the intercept on the A_i/A_{st} axis. A calibration curve passing through the origin has $a = 0$.

After completion of the analysis, the quotient M_i/M_{st} is calculated for the determined A_i/A_{st} ratio using Eq. (53). Mass (m_i) or percentage (P_i) of component i in the sample is then calculated from the equations

$$m_i = \frac{M_i}{M_{st}} \times m_{st} \tag{54}$$

and

$$P_i = \frac{m_i}{m_{as}} \times 100 \tag{55}$$

where m_{st} and m_{as} have the same significance as in Eq. (52).

For this version, it is not necessary to know the correction factors for the individual components, as they are expressed in terms of the slopes of the particular calibration curves.

The precision of the results of the analysis increases with an increasing number of points (i. e. calibration mixtures) used in drawing up the calibration curve (i. e. the greater the response for the given component) and with an increasing number of parallel determinations.

It follows from statistical analysis that increasing the number of points above five has no great effect.

The dependence of the response on the concentration is not linear for some substances; then the general relationship

$$y = ax^b \tag{56}$$

holds for the calibration curve, where $y = A_i/A_{st}$, $x = M_i/M_{st}$ and a and b are constants dependent on the compositions of the sample and internal standard and on the analysis conditions. They can be derived either graphically or by calculation[59, 60].

In the graphical method, the calibration measurement results are plotted on log-log paper and two points with coordinates x_1, y_1 and x_2, y_2 are selected on the straight line obtained. The constants in Eq. (56) are then calculated as follows:

$$b = \frac{\log y_2 - \log y_1}{\log x_2 - \log x_1} \tag{57}$$

$$a = 10^{(\log y_1 - b \cdot \log x_1)} \tag{58}$$

Computer calculation is more precise; the calibration curves are obtained by the least squares method by solving the equation

$$Y = Z + bX \tag{59}$$

where X, Y and Z are the decadic logarithms of quantities x, y and a in Eq. (59), respectively. It then holds for constants a and b that

$$b = \frac{\Sigma X_i Y_i - \Sigma X_i \bar{Y}}{\Sigma X_i^2 - \Sigma X_i \bar{X}} \tag{60}$$

$$Z = Y - bX \tag{61}$$

and

$$a = 10^Z \tag{62}$$

The applicable lifetime of a calibration curve is, of course, limited, as the rods age with an increasing number of analyses. Thus, it is preferable to test the validity of the derived relationships, either by occasionally testing the calibration dependence by analysis of at least two standard mixtures (mostly after 10–15 series of analyses), or by application of the standard mixture to the first two rods in each series of determinations. The remaining eight rods are sufficient for analysis of two samples, as four parallel determinations suffice. This location of the rods in the set is desirable when connecting the detector to a computer which automatically controls the prescribed limits of the applied correction factors in each analysis. This type of control is, of course, useful only when the properties of the first two rods do not differ substantially from those of the remaining eight.

1.3.2.8 Reproducibility in TLC-FID

The reproducibility of the results obtained by TLC-FID analyses has often been discussed. It is apparent that there are a greater number of possible sources of imprecision in TLC than in GLC or HPLC. These include the greater difficulty in controlling both the chromatographic process and the more complicated parameters for FID detection. It thus follows that the reproducibility of TLC--FID is generally poorer than that of GLC[61] (or HPLC), but will at least be comparable with HPLC with spectrophotometric determination, where the results are, moreover, affected by a lack of a standard procedure for detection of the separated zones. The reproducibility of TLC-FID is primarily affected by the following factors:

– homogeneity of the sorbent structure throughout the Chromarod and differences in the chromatographic properties of all the rods in a set

– the amount of sample applied

– the use of standard conditions for chromatography i. e. constant thin layer activity, reproducibility of chamber saturation, maintenance of constant relative humidity in the chamber and a constant composition of the mobile phase, constant development time and constant development temperature

– uniform distribution of the solute around the circumference of the rod, dependent on the sample application technique, on the diffusion coefficient of the solute (s), on the composition of the mobile phase and on the elution time and path

– the volatility of the separated components during drying of the rods after elution and during detection in the FID

– detection parameters, i.e. the distance of the rod from the burner tip, detection rate, hydrogen and air flow-rates and central location of the Chromarod in the FID

– purity of the rod and of the laboratory air, i.e. the intensity of the background signal during the passage of the unoccupied part of the rod through the detector

– the degree of use of the rod (dependent on the first factor, above)

– adjustment of optimal integration parameters

– selection of a method for result evaluation.

The analysis conditions can be optimized on commercial Chromarods to yield reproducibility roughly half that attained for GLC analysis (there is a coefficient of variation of about 2 %. For samples where the amount of analysed component equals 85–95 %, it can be as low as 1 %). In common analytical practice, the average coefficient of variation is about 5 % and can attain 10 % for inexperienced workers.

The problem of reproducibility in the analysis of actual samples will be considered in the following part of this book.

Part 2 Specialized Applications

It is apparent from the published literature that thin-layer chromatography with flame ionization detection TLC-FID is used most extensively for the analysis of simple and complex lipids, especially in biology, in medicine and in the field of vegetable fats and oils. In addition, analyses of pharmaceuticals and natural substances, including foodstuffs, are also important. TLC-FID has been applied since the early 1980s in the analysis of oil and coal products, pesticides, polymers and surfactants.

The following chapters describe the results of the most important papers, with emphasis on analytical conditions, reproducibility and practical use of this new analytical method.

The text is complemented by figures depicting chart recordings made, with a few exceptions, at the Institute of Chemical Technology in Prague, using an Iatroscan TH-10 Mark II instrument. The text also contains tables of elution systems and the pertinent R_F values; these values, of course, only provide guidelines as they depend markedly on maintenance of constant conditions during chromatographic analysis, similarly to classical TLC (see Section 1.2.3). Unless otherwise designated, the composition of the elution and extraction systems are given in volume parts.

2.1 Applications in Biology, Medicine and Pharmacy

2.1.1 Lipids and Related Substances

Lipids form a wide group of substances characterized by the presence of fatty acids ($> C_8$), linked by an ester bond to an alcohol or by an amide bond to a sphingoid base. These can be divided into simple and complex lipids. The former include, primarily, the fatty acid esters with glycerol (fats) and fatty alcohols (waxes) and are usually considered to include ceramides, in which the fatty acid is attached through an amide bond to sphingenine or sphinganine. In complex lipids the free hydroxyl group of the simple lipid is substituted by a

Table 15 Survey of the Most Important Types of Lipids

(a) Simple Lipids

Alcohol	Mol fatty acid/mol alcohol	Further group	Lipid	Abbrev.
glycerol	1	–	monoacylglycerol	MG
glycerol	2	–	diacylglycerol	DG
glycerol	3	–	triacylglycerol	TG
glycerol	2	alkyl	alkyldiacylglycerol	–
glycerol	2	alkenyl	alkenyldiacylglycerol[a]	–
fatty alcohols	1	–	waxes	W
sphinganine[b]	1	–	ceramide	CR
sphingenine[c]	1	–	ceramide	CR
cholesterol[d]	1	–	cholesterol esters	CE

(b) Complex Lipids
(b)(i) Phospholipids
Glycerophospholipids

Basic lipid	Phosphate component	Lipid	Abbrev.
monoacylglycerol	-O-PO(OH)$_2$	lysophosphatidic acid	LPA
monoacylglycerol	lysophosphatidic acid	bis(monoacylglycerol)phosphate[e]	BMP
monoacylglycerol	phosphatidic acid	monoacylglycerol-diacylglycerophosphate[f]	MDP
diacylglycerol	-O-PO(OH)$_2$	phosphatidic acid	PA
diacylglycerol	phosphatidic acid	bis(diacylglycerol)phosphate[g]	BDP
diacylglycerol	phosphocholine	phosphatidylcholine	PC
diacylglycerol	phosphoethanolamine	phosphatidylethanolamine	PE
diacylglycerol	phosphoserine	phosphatidylserine	PS
diacylglycerol	phosphoinositol	phosphatidylinositol	PI
diacylglycerol	phosphoglycerol	phosphatidylglycerol	PG
diacylglycerol	phosphatidylglycerolphosphate	diphosphatidylglycerol[h]	DPG
alkylacylglycerol	phosphocholine	plasmenylcholine	–
alkylacylglycerol	phosphoethanolamine	plasmenylethanolamine[i]	–
alkenylacylglycerol	phosphocholine	plasmanylcholine	–
alkenylacylglycerol	phosphoethanolamine	plasmanylethanolamine	–

Sphingophospholipids

Basic lipid	Saccharide	Lipid	Abbrev.
ceramide	phosphocholine	sphingomyelin	SPM
ceramide	phosphoethanolamine	ceramidephosphoethanolamine	–
ceramide	phosphoinositol+saccharides	phytoglycolipids[j]	PGL
diacylglycerol	galactose, glucose	monoglycosyldiacylglycerol	MGD
diacylglycerol	galactose, glucose	diglycosyldiacylglycerol	DGD
diacylglycerol	6-sulphoglucose	sulphoquinovosyldiacylglycerol	–

(b)(ii) Glycosphingolipids

ceramide	glucose, galactose	monoglycosylceramides[k]	MGCR
ceramide	polysaccharide	polyglycosylceramides	–
ceramide	fucose	fucolipids	–
ceramide	globotriose	globosides	–
ceramide	globotetrose	globosides	–
ceramide	mucotriose	mucoglycolipids	–
ceramide	lactotriose	lactoglycolipids	–
glycosylceramide	sialic acid[l]	sialoglycosphingolipids[m]	–

a neutral plasmalogen; b 2S,3R-2-amino-1,3-octadecandiol; c 2S,3R,4E-2-amino-4-octadecen-1,3-diol; d belongs among the steroids, and is often, but incorrectly, included among neutral lipids; e better known incorrectly as lysobisphosphatidic acid; f semilysobisphosphatidic acid; g bisphosphatidic acid; h cardiolipin; i plasmalogen; j contains phosphoinositol, to which other saccharides such as mannose, glucosamine, glucuronic acid, etc. are bonded through a glucoside bond; k cerebrosides; l N-acetylneuraminic acid; m gangliosides.

further polar component, containing a phosphate group (phospholipids) or a saccharide (glycolipids), etc. Table 15 lists basic information on the most important types of lipids. More detailed information on nomenclature and structure can be found in the literature[62—65].

Knowledge of the importance and the role of lipids in cell metabolism, immunology, diagnosis of civilization-induced diseases and nutrition is closely related to the discovery of new effective analytical separation and identification methods. In addition to the classical methods of chemical and enzymatic analysis[66,67] and separation of lipid classes by column chromatography on silica gel, DEAE cellulose, carboxymethylcellulose and other substances[64], these methods consist, primarily, of gas chromatography of intact lipids[68—71], TLC and HPLC[64].

Gas chromatography is now an almost irreplaceable tool in the separation of simple lipids according to their carbon number, but cannot be used to separate phospholipids and other polar lipids. Thus, combining gas chromatography with the TLC-FID method is a logical step in obtaining additional information on the composition and content of lipids in biological materials.

The first TLC-FID application in lipid analysis was published at the beginning of the seventies by the Japanese authors Kawai[72], Nakano[73], Tokunaga[74], Ueda[75] and Inoue[76]. This work was expanded in particular by Vandamme[77,78], Herslöf[79], van Tournout[80], Bradley[48], etc.[61,81—83].

2.1.1.1 Isolation of Lipids from Biological Materials

The first step in the analysis consists of extracting the lipid from the biological material. Simple lipids are soluble in a wide variety of polar and non-polar solvents, such as n-hexane, diethyl ether, benzene, chloroform, ethyl acetate and acetone. The two main complex lipid groups, i. e. phospholipids and glycolipids, differ in their solubilities in acetone and ethyl acetate. Phospholipids do not dissolve in these solvents (except for phosphatidic acid and some other acidic phospholipids), while glycolipids are soluble. Non-polar compounds such as triacylglycerols and cholesterol esters are not soluble in methanol. Their solubility in alcohols increases with the molecular weight of the alcohol, so that most non-polar lipids are readily soluble in n-butanol. Most lipids are insoluble in water, but a considerable number of polar lipids, e. g. lysophosphatidic acids and especially gangliosides, are soluble in aqueous solutions of the lower alcohols. An ideal solvent for extracting lipids from animal and plant tissues should be suitably polar to be able to overcome association between the lipid molecules and proteins in both cell membranes and lipoproteins, but should not react chemically with the lipids[64,84].

Isolation of lipids from tissues is based, primarily, on the character of the biological material. It is relatively simple to extract lipids from liquid samples, such as serum or plasma, obtained from the blood after removal of blood clots or erythrocytes. The isolation of lipids from animal and plant tissue is more complicated and homogenization is often necessary for these samples.

Extraction using a five-stage extraction procedure is considered ideal. This involves double extraction using a mixture of chloroform and methanol (2:1, called the Folch procedure[85]), followed by single extractions with chloroform-methanol (1:2), chloroform-methanol (7:1, saturated with ammonia), chloroform-methanol-acetic acid-water (4:2:1:0.5) and finally chloroform--methanol-concentrated hydrochloric acid (200:100:1)[86]. In practice, however, the first two systems are more commonly[87] used and, in the presence of polyphosphoinositides, which can be found especially in brain tissues and yeast, the third system consists of the hydrochloric acid system[88]. If the lipid fraction contains large amounts of lysophospholipids, it has been recommended that the chloroform-methanol mixture be replaced by n-butanol saturated with water[89].

Sometimes viscous residues are formed during extraction with a chloroform--methanol (2:1) mixture and cannot be dissolved even at high homogenizer speeds. It is then preferable to add the chloroform fraction (ca. 20 ml per gram of sample) after prior homogenization in methanol (10 ml g^{-1})[90].

Special attention should be paid to the isolation of lipids from plant tissues containing phospholipase D, an enzyme that catalyses the reaction of phospholipids with methanol to produce undesirable artefacts, primarily phosphatidylmethanol, which can be easily confused in TLC separation with bis(diacylglycerol)phosphate (bisphosphatidic acid). Phospholipase can be quite easily deactivated by steaming the tissue (ground seeds) for 15 min prior to extraction[91], or by substituting methanol by isopropyl alcohol, which also acts as a deactivator[92]. Because of its lower toxicity, a mixture of isopropyl alcohol with n-hexane has also been recommended for the isolation of lipids from animal tissue[93]. However, gangliosides are incompletely extracted by this system[64].

Larger samples (about 100 g) are usually extracted using the procedure recommended by Bligh and Dyer[94] (in the Christie modification[64]). Here, the amounts of chloroform and methanol (1:2) are adjusted in dependence on the amount of water in the sample so that a single -phase system is formed with a chloroform-methanol-water ratio of 5:10:4. Tissues with a high lipid content, such as adipose tissue or oil seeds are first extracted with diethyl ether or chloroform and then the isolation is completed by the above procedure[64].

The raw extract is usually contaminated to a greater or lesser extent by saccharides, urea, amino acids, proteins and salts, which complicate the chromatographic separation of lipids. Most of these substances can be removed by washing the extract with water or, preferably, by using an aqueous solution of potassium chloride (0.9 %, in an amount equal to approximately one third of the overall extraction volume)[64]. A more modern purification procedure involves distribution chromatography of the extract in a chloroform-methanol-water system on Sephadex G 25[95, 96]. At a higher glycolipid content (especially gangliosides), which are extracted into the aqueous phase, it is preferable to evaporate the extract in a vacuum evaporator and redissolve it in a chloroform-n-hexane (1 : 3) mixture[97] and then centrifuge or filter the solution obtained.

A brief description will now be given of the preparation of the extract prior to analysis by the TLC-FID method.

Lipid chromatography is carried out using 50 µl to 0.5 ml of serum or plasma which is extracted with a 20–30-fold volume of solvent; the non-lipid fraction is removed by filtration and the extract is washed with either water or a salt solution. The chloroform and methanol are then removed at low temperature and reduced pressure[48, 77, 80, 82, 98].

A very simple procedure is described in the manual supplied with the Iatroscan instrument[99]: 0.5 ml of plasma or heparinized blood is shaken for 30 s with 10 ml of a mixture of chloroform and methanol (2 : 1), filtered and the extract is washed with 2 ml water. The water is removed by aspiration after centrifuging for 10 min at 2500 rev min^{-1} and the organic phase is evaporated at 40 °C on a water bath. The whole operation takes less than 1 h. We have found, however, that the lipid extraction is not complete in this procedure and washing with water alone can lead to loss of part of the free fatty acids. It is preferable to extract for 15 min in a separatory funnel and wash with a 2 % calcium chloride or 0.9 % potassium chloride solution.

The preparation of lipid extracts from other materials is more difficult, especially in the first phase (isolation of erythrocyte membranes, hepatocyte membranes, etc.). The subsequent extraction is similar to the above procedure. For example, erythrocyte membrane lipids (erythrocyte ghost) prepared by the Dodge[100] or Takemoto[101] method are first extracted with methanol or isopropyl alcohol and then homogenized. The required amount of chloroform is added (ca. 1 vol. part per part of alcohol) and the extraction is carried out for ca. 1 h with occasional mixing. Water is then added (about 0.5 parts) and the mixture is thoroughly mixed and separated in a centrifuge. The chloroform layer is evaporated at low temperature (20–40 °C)[102, 103]. Lipids are isolated from brain tissue by double extraction with intense stirring. For example, cerebral white matter is first extracted with 20 volumes of a chloroform-methanol (2 : 1) mixture in a glass homogenizer. The homogenized mixture is then centrifuged and the precipitate is again extracted in the same way with chloroform-methanol (1 : 1). The combined extracts are evaporated and dissolved in chloroform; the solution obtained is washed with 0.2 volumes of water[104].

Lipids in amniotic fluid are first prepared by extracting the lipids in the prescribed manner[105] and then analysing them directly by TLC-FID. Prior to the actual analysis, separation into fractions soluble and insoluble in acetone can be carried out[106].

2.1.1.2 Chromatography on Chromarods

The extract is dried and the residue is then dissolved in chloroform or chloroform-methanol (2:1 to 4:1) to give a 1–2 % solution. An amount of 0.5–2 µl is applied to Chromarods S or SII. Chromarods A are used for lipid analysis only rarely.

The rods are then developed in one of three ways:

– separation of neutral lipids from the remaining phospholipids by single-step elution in a low polarity system, e. g. in a mixture of n-hexane with diethyl ether

– subsequent development with two or three solvent systems with different polarities; all the simple and complex lipids are separated along the whole length of the rod

– stepwise development using selective pyrolysis of neutral lipids; the first elution in a less polar system separates the neutral lipids which are then detected in the FID. Phospholipids and glycolipids, where present, are then separated in the reactivated part of the layer using a more polar system.

The first procedure is simple and rapid, but yields information only on the contents of simple lipids in the sample, i. e. the contents of fatty acids, triacylglycerols (and possibly also mono- and diacylglycerols) and esterified and free cholesterol. Phospholipids are not separated and remain at the origin. For some clinical purposes, this information is sufficient, e. g. in control of the levels of cholesterol and triacylglycerols in the blood of patients suffering from hypercholesterolemia or hypertriacylglycerolemia[61]. Up to six analytical cycles (with 10 rods per cycle) can be carried out per working shift, i. e. three repeated lipid analyses on 18 samples.

The second elution procedure is more tedious but leads to complete separation of all the lipid classes in the sample, i. e. phospholipids, cholesterol along with its esters with fatty acids, triacylglycerols and free fatty acids. When the sample contains a large number of components (10–15), partial overlapping of the separated zones can occur, leading to a loss of precision and reproducibility. The quality of the separation depends on the activity of the thin layers on maintenance of optimal elution pathways in the individual systems. Far fewer analyses of this type can be carried out in a single day (two or three cycles, i. e. six to nine samples).

In stepwise development with selective scanning, the area on the rod available for chromatography is almost twice as large as in the previous two procedures. The peak density in the chromatogram is thus proportionately smaller. The greater wear on the layer by repeated pyrolysis, prolonged stay in the flame (by about 30–40 %) and problems involved in quantitative evaluation of two scans on a single rod are the disadvantages of this approach. Examples of application of the two procedures described are given in Figures 17 and 23.

Solvent systems for single-step chromatography of simple lipids are based on combinations of n-hexane or petroleum ether with diethyl ether, with a few exceptions, and sometimes they also contain small amounts of low molecular weight organic acids, especially formic (called the HDF system). Table 16 lists the solvent system compositions employed and the corresponding R_F values.

It is apparent from the table that the type of elution system affects primarily the separation of fatty acids and triacylglycerols.

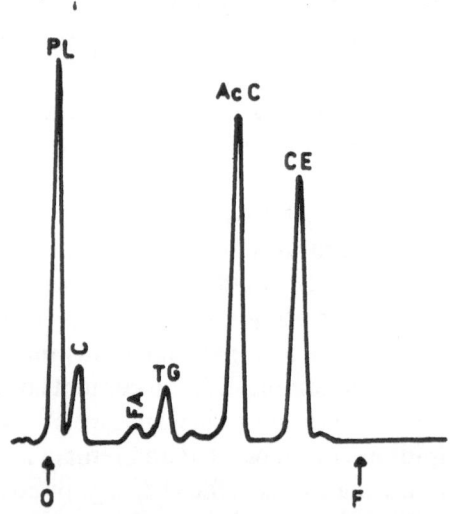

Fig. 23. Analysis of lipids in blood plasma of a healthy individual. Single-step elution with n-hexane-diethyl ether-formic acid 92 : 7.85 : 0.15. Chromarods S II, scanning speed 0.42 cm s^{-1}, H$_2$ flow-rate 180 ml s^{-1}, air 2.1 l min^{-1}. Abbreviations as in Figure 17; AcC cholesterol acetate (internal standard); F front of the chromatogram

Table 16 Systems for Single-step Elution of Lipids and the R_F Values for the Separated Components
(a) System compositions (% vol.)

No.	n-Hexane	Petroleum ether	Diethyl ether	Formic acid	Acetic acid	Chloro-form	Refe-rence
1	90		10				76,107
2	85		15				77,78
3	85.7		14.3				108
4	83.9		16.0	0.1			36
5	79.2		19.8		1.0		79
6		96.0	3.0	1.0			109
7		96.0	3.0		1.0		36
8		84.9	15.0	0.1			80
9[a]					0.1	8.0	110
10[b]				0.5		29.8	111
11	89.8		10	0.2			111

a plus 91.9 % 1,2-dichlorethane; b plus 69.7 % benzene.

(b) R_F values

The meaning of the abbreviations is given in the List of Symbols

System no.	Chroma-rod	PL	C	FA	TG	CE	Internal standard
1	S c	0.06	0.12	–	0.40	0.80	0.50[e]
	S d	0.05	0.12	0.17	0.42	0.70	0.54[e]
2	S	0.08	0.18	0.31	0.53	0.71	–
3	A	–	0.07	–	0.45	0.76	–
4[f]	S II	0.07	0.24	0.33	0.50	0.64	0.73[g]
5	S	–	0.42	0.60	0.68	–	0.17[h]
6	S	0.06	0.17	0.43	0.33	0.66	–
7	S	–	0.12	0.50	0.28	0.67	–
8	S	0.07	0.16	0.35	0.47	0.70	0.25[i]
9	S	0.05	0.35	0.46	0.59	0.80	–
10	S II	0.00	0.29	0.59	0.69	–	–
11	A	0.00	0.16	0.28	0.48	0.81	–

c reference 76; d reference 107; e cholesterol acetate; f double development, 1,2-DG and 1,3-DG (R_F 0.15 and 0.24, resp.) also separated; g fatty acid methyl ester; h chimyl alcohol; i octadecanol.

Addition of organic acids to the system (formic or acetic) increases the R_F value of almost all the components and improves the separation of the fatty acid zone. In systems without acids, this zone is often broad and deformed.

Phospholipids are not separated in these systems and practically do not migrate. If the sample contains monoacylglycerol, it also remains close to the start. The positional isomers of diacylglycerol are separated quite well on Chromarods S II (system no. 4), but the 1,3-isomer forms a critical pair with free cholesterol.

A mixture of chloroform, dichloroethane and acetic acid (no. 9) differs markedly from the others in the series. This unusual combination increases the mobility of cholesterol and the fatty acids, but developing times are longer.

A mixture of benzene, chloroform and formic acid is also rather unusual; this mixture is used for the separation of lipids from the microsomes of guinea pig liver cells (system no. 10, Table 16)[111].

Kramer et al.[112] considered the effect of the n-hexane-diethyl ether-organic acid solvent system on the separation of simple lipids. In contrast to HPTLC on silica gel layers (of the HP-K type), where the fatty acid zone on the chromatogram has a lower R_F value than triacylglycerol at all diethyl ether levels, the positions of these lipids on Chromarods S depends on the solvent system composition (Table 17). In solvents without organic acids, the R_F value for fatty acids is low and triacylglycerol lies between the acid and the ester of cholesterol (system no. 1, Table 16).

The mobility of fatty acids is also low on Chromarods A even when the solvent system contains formic acid (system no. 11, Table 16). Chromarods A are less suitable for lipid separation, as the separated zones are much broader than on silica gel.

The migration of the more polar components on Chromarods is much faster than on planar thin silica gel layers. The chromatographic properties of the two layers are different although they are prepared from the same adsorbent. This difference is apparently a result of the presence of the carrier frit layer, which can participate in the chromatographic separation as a result of its capillary system which has high surface activity[112].

An increase in the diethyl ether content in systems containing 1 % by volume

Table 17 R_F Values of Simple Lipids on Chromarods S and HPTLC (Type HP-K) in Systems of the n-Hexane – Diethyl Ether – Formic Acid Type (or acetic or propionic acids). Derived from results presented by Kramer et al.[112]

(a) Comparison of R_F values for HPTLC and Chromarods

Vol. parts n-hexane	Diethyl ether	Formic acid	C	FA	TG	ME[a]	CE	Type of layer
95	5	0.1	0.01	0.00	0.05	0.26	0.44	HPTLC
97	3	0.1	0.14	0.22	0.18	0.43	0.58	CH-S
95	5	1.0	0.03	0.00	0.09	0.36	0.58	HPTLC
95	5	1.0	0.20	0.49	0.39	0.59	0.74	CH-S
85	15	0.1	0.04	0.09	0.32	0.53	0.73	HPTLC
85	15	0.1	0.23	0.47	0.56	0.63	0.74	CH-S
85	15	1.0	0.05	0.17	0.31	0.51	0.71	HPTLC
85	15	1.0	0.38	0.65	0.68	0.68	0.78	CH-S

(b) R_F values on Chromarods S in dependence on the diethyl ether content; the formic acid content is constant (1 part by volume), the content of n-hexane (parts by volume) = 100 − Et$_2$O content.

Vol. parts diethyl ether	C	FA	TG	ME[a]	CE
3	0.15	0.43	0.26	0.51	0.68
5	0.20	0.49	0.39	0.59	0.74
10	0.30	0.60	0.60	0.66	0.78
15	0.38	0.65	0.68	0.68	0.78
20	0.38	0.67	0.70	0.70	0.78

(c) R_F values on Chromarods S in dependence on the type of organic acid in the n-hexane – diethyl ether – acid system (85:15:0.1).

Acid	C	FA	TG	ME[a]	CE
formic	0,23	0.47	0.56	0.63	0.74
acatic	0.20	0.50	0.56	0.67	0.79
propionic	0.23	0.55	0.64	0.74	0.86

(d) The effect of ageing of Chromarods S on the R_F values in the n-hexane – diethyl ether (97:3, v/v) – formic acid system

Vol. parts formic acid	Rods[b]	C	FA	TG	ME[a]	CE
0.1	N	0.14	0.22	0.18	0.43	0.58
0.1	S	0.14	0.22	0.18	0.38	0.50
0.5	N	0.15	0.37	0.20	0.47	0.62
0.5	S	0.15	0.37	0.20	0.40	0.52
1.0	N	0.15	0.43	0.26	0.51	0.68
1.0	S	0.15	0.43	0.26	0.44	0.53

a fatty acid methyl ester; b N – new set of Chromarods S; S set of Chromarods S used for 20–30 analyses.

of organic acid leads to gradual equalization of the R_F values of fatty acids, triacylglycerol and methyl ester. Thus diethyl ether primarily increases the mobility of triacylglycerol (Table 17). The separation of all the components can be improved by decreasing the content of organic acid to 0.1 % with an increase in the diethyl ether content to 15 % (Table 17). It can also be seen from the table that the mobility of all the components (except cholesterol) increases with an increase in the molecular weight of the organic acid in the system.

The effect of rod ageing on the migration of the less mobile components is also interesting (Table 17d). This unfavourable effect may be a result of a decrease

Table 18 R_F Values for Simple Lipids on Chromarods S Developed by Stepwise Elution
(a) System composition (volume parts)

No.		n-Hexane	Petroleum ether	Diethyl ether	Acetic acid	Ethanol	2-Methyl-propane--1-ol	Methanol	Reference
1	(a)	100			1				48
	(b)	100			1		0.5		
2	(a)			75		25			
	(b)		90	10	1				114
	(c)							100	

1(a): developed for 15 min; 1(b): 20 min; 2(a): to 2 cm; 2(b): to 10 cm; 2(c): twice to 4 cm.

(b) R_F values

System no.	PL	C	FA	TG	CE	IS
1	0.00	0.20	0.40	0.60	0.76	0.11*
2**	0.14	0.30	–	0.50	0.76	–

 * umbelliferylpalmitate
** plus proteins, R_F 0.00.

in the porosity of the older rods as a result of pore clogging and can generally be eliminated by occasionally rinsing the rods in dilute hydrochloric acid.

Kramer et al.[112] confirmed the results of Parrish and Ackman[113], who also pointed out the high dependence of the mobility of fatty acids in HDF type systems on the relative humidity of the laboratory atmosphere. The reproducibility of the R_F values of the fatty acids can be greatly improved by conditioning the rods for about 10 min in a chamber with constant humidity prior to the elution.

Problems connected with the sensitivity of the separation of triacylglycerols and fatty acids using systems of the n-hexane-diethyl ether type were considered by Bradley et al.[48], who used Chromarods S and stepwise elution with n-hexane-acetic acid (100:1, 15 min.), and n-hexane-acetic acid-2-methylpropan-1-ol (100:1:0.5, 20 min)[48]. It can be seen from Table 18 that the separation of all the lipids on the rod is good and that triacylglycerol occupies the "normal" position between the fatty acid and the cholesterol ester.

Mills et al.[114] also employed stepwise elution in the separation of individual lipids from the sum of the phospholipids and proteins in the lipoprotein fraction of HDL and LDL, obtained by ultracentrifuging serum (system no. 2, Table 18). In the first elution lipids are separated from phospholipids and proteins on a short path length (2 cm) using a mixture of ethanol and diethyl ether. The lipids are then separated into their individual components in the next elution using the usual system of petroleum ether, diethyl ether and acetic acid. The rod is then developed twice with methanol (to 4 cm from the start), separating the polar lipids from the proteins which remain at the start. Prior to FID detection, the rod is immersed in distilled water for 5 min, to remove residues of salts that remain in the analysed sample of lipoprotein after purification by dialysis and prior to ultracentrifugation. Water is then removed by drying for 20 min at 100 °C. The entire analytical operation lasts 80 min, and 10 measurements require 100–200 µl of sample.

As mentioned in the introduction to this section, neutral lipids can be readily separated using gas chromatography which, however, cannot be employed to separate phospholipids. It is thus not surprising that one of the important applications of TLC-FID is the chromatography of polar lipids, whose physiological importance is the subject of an ever increasing number of papers. If simultaneous chromatography of phospholipids and neutral lipids on a single layer is not involved, then single-step elution is sufficient; a polar solvent system distributes phospholipids over the whole length of the rod. The neutral lipids remain at the front or close to it (Table 19).

Recently, simple and complex lipids have been separated simultaneously on a single rod, either using the partial scanning method mentioned in Section 1.2.3 (see Figure 17), or, more often, using a combination of multiple and stepwise elution. The latter yields a single precise chromatogram that permits quan-

Table 19 Compositions of Solvent Systems for Single-step Elution of Phospholipids and R_F Values
(a) System composition (vol. %)

System no.	Chloroform	Methanol	Water	Ammonium hydroxide	Reference
1	74.1	23.1	2.8		106
2	70.6	27.2		2.2	111
3	70.0	26.2	3.8		114
4	67.8	29.7	2.5		115
5	67.5	29.5	3.0		116
6	66.7	29.2	4.1		78
7	65.9	31.1	3.0		36.117
8	64.6	32.2	3.2		101.118
9	64.2	32.1	3.7		107.119
10	64.0	32.0	4.0		111
11	63.2	31.6		5.2	79
12	61.3	35.3	3.4		61
13	58.9	39.3	1.7	0.1	103

(b) R_F values

System no.	Chroma-rods	LPC	SM	PC	PS	PE	NL[f]	Note
1	S	0.15	0.25	0.46	–	–	–	
2	S II	–	0.07	0.16	–	0.36	0.85	a
3	S	0.12	0.25	0.49	0.62	0.72	–	b
4	S	0.22	0.27	0.36	0.46	0.70	0.86	c
5	S	–	0.18	0.28	0.44	0.62	0.76	
6	S	0.17	0.24	0.38	–	0.83	0.83	
7	S II	0.17	0.30	0.40	0.61	0.70	0.80	d
8	S	–	0.16	0.26	0.40	0.53	0.65	ref. 101
8	S II	–	0.14	0.27	0.53	0.66	0.83	ref. 111
8	A	0.12	–	0.26	0.41	0.66	–	ref. 118
9	S	–	0.14	0.26	0.46	0.70	–	e
10	S II	0.12	0.22	0.34	–	–	0.83	
11	S	–	–	0.37	–	–	0.62	
12	S II	0.16	0.24	0.38	–	0.73	0.86	
13	–	–	0.15	0.23	0.32	0.47	0.77	

a plus two additional substances with R_F 0.74 and 0.57 (probably glucosides); b plus ceramides and cerebrosides (R_F 0.88), R_F identical to that for PS; c plus PI (R_F 0.58); d plus PI (R_F 0.52); e plus PI (R_F 0.58); f neutral lipids.

titative evaluation with a single internal standard. Obviously, the large number of components in the sample (more than 10) places stringent requirements on selection of a suitable elution agent(s), determination of the optimum path length of the individual elutions and, especially, maintenance of constant conditions during the chromatography. Here, saturated elution chambers that can be presaturated are especially useful (Section 1.2.2, Figure 8). Table 20 surveys the elution systems used together with the results obtained.

It can be seen from these two tables that most elution systems contain a mixture of chloroform, methanol and water, i. e. a system used in the 1950s for the chromatography of phospholipids on paper impregnated with water glass[120], that has also been used in classical TLC on silica gel layers[121].

These systems are useful for separating neutral ampholytic phospholipids. However, they are less useful for acid lipids, such as phosphatidic acid, phosphatidylserine and phosphatidylinositol and other substances forming broad zones during migration in the layer that tail or have R_F values dependent on concentration[122].

In TLC-FID on planar silica gel layers, this problem is solved either by addition of ammonia, which decreases the R_F value of acid phospholipids and limits tailing or, on the other hand, by acidification of the elution solvents using acetic acid, which increases the mobility of all the acidic components and sometimes markedly increases the sharpness of the separation.

So far, basic and acid systems have found limited application in TLC-FID. The few examples listed in Tables 18 and 19 can be considered as preliminary but not very successful experiments. The addition of ammonium hydroxide has practically no effect on the separation of amphoteric phospholipids (phosphatidylcholine (PC), phosphatidylethanolamine (PE) and sphingomyelin (SM)), but leads to the separation of some acid phospholipids (phosphatidic acid (PA) and phosphatidylserine (PS)) into two peaks. One of these lies close to the start and has an R_F value similar to that lysophosphatidylcholine (LPC). The other lies in the first third of the chromatogram and almost coincides with the phosphatidylcholine peak (see Table 20, systems 5–8). This is apparently connected with the partial neutralization of the acid groups of the given glycerophospholipid by ammonia, so that one substance (PA, PS) yields two types salts (acid and neutral) with different mobilities. This could probably be prevented by increasing the content of ammonium hydroxide in the system; however, the fatty acid zones would then coincide (as ammonium soap) with free cholesterol. The presence of ammonia in the system is also unfavourably reflected in the gradual decrease in the separation capacity of silica gel Chromarods, probably as a result of a decrease in the concentration of active adsorbent by washing out of the ammonium silicates formed. In addition, systems with ammonium hydroxide usually yield a higher noise level during scanning.

In systems of the chloroform-methanol type, acetic acid usually increases the mobility of phosphatidic acid and phosphatidylserine and free fatty acids, which can coincide with triacylglycerol. It can be removed from the elution layer only by drying at elevated temperatures.

It follows, therefore, that an elution system consisting of a mixture of chloroform, methanol and water remains most suitable and yields quite good separation of the most important phospholipids (LPC, SM, PC, PI, PE and DPG), especially on Chromarods S II. Lysophosphatidylethanolamine (LPE) forms a critical pair with phosphatidylserine and diphosphatidylglycerol (DPG) with monoacylglycerol and some glycolipids. The optimal chloroform-methanol ratio is roughly 2:1. A decrease in the methanol content inhibits the separation

Table 20 Solvent Systems for Stepwise Development of Mixtures of Simple and Complex Lipids on the One Rod and the R_f Values

(a) Solvent system composition

System no.	Chloroform	Methanol	Water	Benzene	n-Hexane	Diethyl ether	Formic acid	Acetone	Ethylacetate	Ammonium hydroxide	Acetic acid	Reference
1	70	26.2	3.8									35
2 (a)	65	30	5									36
(b)			1.3	92			0.1	4	2.6			117
(c)					92,9	7.0	0.1					
3 (a)	67.8	29.7	2.5									36
(b)			1.3	92			0.1	4	2.6			117
(c)					92.9	7.0	0.1					
4 (a)	64.0	32.0	4.0									99
(b)					89.9	10.0	0.1					111
5 (a)	68.0	28.0	3.9							0.1		36
(b)			1.3	92			0.1	4	2.6			117
(c)					92.9	7.0	0.1					
6 (a)	63.0	31.0	4.0							2.0		36
(b)			1.3	92			0.1	4	2.6			117
(c)					89.9	10.0	0.1					
7 (a)	73.0	21.0	4.0							2.0		36
(b)			1.3	92			0.1	4	2.6			117
(c)					89.9	10.0	0.1					
8 (a)	69.7	26.0	2.6							1.7		36
(b)	77.0	19.2	1.9								1.9	117
(c)					89.9	10.0	0.1					
9 (a)	67.0	28.0	3.0								2.0	36
(b)					89.9	10.0	0.1					117
10 (a)	69.9	21.0	–						7.0		2.1	36
(b)					89.9	10.0	0.1					117

(b) R_F values

System no.	Chroma-rods	No. of developments	Elution path cm (min)	LPC	SM	PC	PS	PI	PE	DPG	MG	C	FA	TG	IS	CE
1*	S II	2	(60)	0.18	0.30	0.44	–	–	0.77	–	–	–	–	–	–	–
**	S II	2	(40)	0.24	0.35	0.45	–	–	–	–	–	–	–	–	–	–
2 (a)(b)(c)	S	1 / 1 / 1	5.5 / 7 / 10	0.09	0.11	0.15	–	–	0.22	–	–	0.32	0.42	0.49	0.60	0.73
2 (a)(b)(c)	S	2 / 1 / 1	5.5 / 7 / 10	0.10	0.15	0.21	–	–	0.27	–	–	0.35	0-44	0.52	0.61	0.75
2 (a)(b)(c)	S II	1 / 1 / 1	5.5 / 7 / 10	0.07	0.13	0.20	0.24	–	0.27	–	0.36	0.40	0.43	0.56	0.69	0.82
3 (a)(b)(c)	S II	1 / 1 / 1	6.5 / 7 / 10	0.13	0.18	0.23	–	–	0.38	–	0.53	0.61	0.68	0.73	0.78	0.82
4 (a)(b)	S II	2 / 1	5 / 10	0.12	0.19	0.27	0.33	0.40	0.43	–	–	0.50	0.61	0.67	–	0.76
5 ***(a)(b)(c)	S II	2 / 1 / 1	5 / 7 / 10	0.10	0.15	0.24	0.07 0.32	–	0.40	–	–	0.55	0.62	0.67	0.73	0.78

98

System no.	Chroma-rods	No. of developments	Elution path cm (min)	LPC	SM	PC	PS	PI	PE	DPG	MG	C	FA	TG	IS	CE
6 (a)	S II	1	6													.
(b)		1	7	0.13	0.27	0.38	0.32	–	0.45	0.53	0.49	0.56	0.60	0.68	0.71	0.78
(c)		1	10													
7 (a)	S II	1	6													
(b)		1	7	0.11	0.13	0.16	0.11	–	0.19	0.25	0.43	0.50	0.58	0.69	0.73	0.80
(c)		1	10				0.19									
8 (a)	S II	1	6													
(b)		1	7	–	0.17	0.25	0.18	–	0.35	0.38	–	0.43	0.55	0.55	0.62	0.72
(c)		1	10				0.25									
9 (a)	S II	1	5													
(b)		1	10	0.08	0.16	0.23	0.37	–	0.37	–	–	0.42	0.54	0.58	0.66	0.73
10 (a)	S II	2	5													
(b)		1	10	0.11	0.15	0.27	0.30	–	0.30	0.39	–	0.43	0.55	0.58	0.66	0.77

* plus LPE (R_F 0.56); ** plus cerebrosides (R_F 0.89); *** plus PA (R_F 0.32 and 0.07).

in the upper and central parts of the layer, so that the differences in the R_F values of diphosphatidylglycerol, phosphatidylethanolamine, phosphatidylserine and phosphatidylinositol (PI) gradually decrease.

Replacement of part of the methanol by isopropyl alcohol or n-propanol leads to a decrease in the separation of sphingomyelin from phosphatidylcholine, phosphatidylserine from phosphatidylethanolamine and triacylglycerol forms a critical pair with cholesterol acetate. In addition, the use of these high molecular weight alcohols increases the elution time and the background noise of the layer during FID scanning[117]. The water content in the system primarily affects the separation of the more polar components and depends on the methanol content. At 30 % methanol, the system should contain 3–4 % water. An important factor in the separation of phospholipids can be the free water content in the activated layers. This depends especially on the humidity of the laboratory atmosphere and possibly also on the conditioning time of the rods in a chamber with a humid atmosphere prior to the actual elution. It follows from Table 21 that increasing hydration of the layer leads to a decrease in the migration of phosphatidylserine[116].

Table 21 Dependence of the R_F Values of Phospholipids and Cholesterol on the Degree of Hydration System 5, Table 18[116].

Degree of hydration	SM	PC	PS	PE	C
low	0.18	0.28	0.55	0.64	0.76
medium	0.18	0.28	0.44	0.62	0.76
high	0.18	0.28	0.35	0.61	0.76

Takemoto[101] studied the effect of humidity on Chromarod separation. The composition of the elution system was adjusted in dependence on the humidity of the laboratory atmosphere. If poor separation was found in the upper part of the rod for the given system (high degree of hydration), chloroform was added to the system; when lysophosphatidylcholine, sphingomyelin and phosphatidylcholine were poorly separated near the start, then a system with a higher methanol content was used. He also found that sharpness of the separation is increased by shortening the elution path to 7–8 cm, corresponding to 18–22 min elution in system 8, Table 19. It is preferable to shift the start to a point 4 cm from the bottom edge of the layer and for the chamber to contain 90 ml of solvent, i. e. 20 ml more than usual. The Takemoto procedure has been found to be especially useful for single-step elution chromatography of phospholipids on Chromarods S II. In stepwise or repeated development, the quality of the separation is less dependent on the atmospheric humidity. Of the systems in Table 20, nos. 2 and 3 were found to be most useful. On a 5–6 cm pathway, the chloroform-methanol-water system separates phospholipids and neutral lipids (except for monoacylglycerol) remain almost at the front. Second elution (using a benzene-acetone-ethyl acetate-water-formic acid system) improves the final separation of monoacylglycerol from phosphatidylethanolamine and diphosphatidylglycerol, and also from free cholesterol. The third system (n-hexane-diethyl ether-formic acid) separates both simple lipids and the internal standard (cholesterol acetate) in the upper part of the rod. Free fatty acids can be separated from triacylglycerol by maintaining the formic acid content in the last elution system at the level necessary to prevent tailing of fatty acids (i. e. ca. 0.05 to 0.1 %). If the analysed sample contains diacylglycerols, then they have an R_F value almost identical to that for cholesterol. This critical pair cannot be separated by the systems in Tables 19 and 20.

The lower separation capacity of S rods compared to S II appears in incomplete separation of lysophosphatidylcholine from sphingomyelin. These compounds can be readily separated by repeated elution in a chloroform-methanol-water system (see separation on S II rods using system no. 2 in Table 20).

Chromarods A have been used only rarely in the separation of phospholipids[118]. The separation of simple mixtures on these rods is adequate (system no. 8, Table 19), but the separated zones are much broader than on S or S II rods.

It can be seen from the chromatograms of natural lipids and model mixtures, prepared from the derivatives of saturated acids (e. g. palmitic), that the model mixture is separated more efficiently. This is mainly because lipids derived from biological materials consist of a wide range of derivatives of saturated and unsaturated acids with various numbers of double bonds, each having a different mobility. The separation of such mixtures can sometimes be improved considerably by hydrogenation prior to application by aerating the sample solution (containing about 10 % by weight of platinum oxide) with hydrogen for 30 min and separating the catalyst in a centrifuge. However, the broad zones of some components can arise from the presence of partly oxidized bonded fatty acids. Hydrogenation is of little use under these circumstances.

Very little data is available on the analysis of glycolipids by TLC-FID, primarily because of problems with separation of more complicated mixtures of native glycolipids by adsorption TLC. Such mixtures contain a large number of compounds with similar composition, differing in the stereoconfigurations of their saccharide residues, which practically cannot be separated without prior formation of less polar derivatives. In addition, the presence of sphingoglycolipids with the same saccharide residue is also often a complicating factor. These compounds may also differ in the length of the fatty acid in the ceramide part of the molecule. These lipids are then separated into two or more zones[65], or yield broad irregular peaks.

A mixture of chloroform, methanol and water is a common solvent system for the TLC of glycolipids, i. e. systems similar to those for the separation of phospholipids. The ratio of chloroform to methanol (and water) depends on the length of the saccharide chain in the glycolipid. The longer this chain is, the greater the content of polar components required in the solvent system[65]. This type of system (chloroform-methanol-water, 72:27:1) has been used for the TLC-FID of globosides (R_F 0.20), trihexosylceramides (0.45), dihexosyl-ceramides (0.61) and galactocerebrosides (0.71) on Chromarods S II[111]. A similar system (78.4:19.6:2) can be used to separate glycolipids, phospholipids and simple lipids from an extract of thermoacidophilic bacteria derived from sulphur-containing spring water[81, 123]. Phospholipids form a doublet with R_F 0.08 to 0.24, glycolipids and simple lipids yield singlets with R_F values of 0.50 and 0.72. The mobility of the phospholipids compared to the glycolipids can be suppressed by addition of acetone to the system, e. g. using chloroform-acetone--water-acetic acid (10:90:3:2). The phospholipids remain at the start and diglucosyldiacylglycerols and monoglucosyldiacylglycerols form two irregular peaks with R_F values of 0.37 and 0.70[124].

TLC-FID can also be used for the determination of the molar ratio between saccharides, sphingoid bases and fatty acid in the simpler glycolipids, e. g. monoglycosyl or diglycosylceramides[125]. The sample of pure glycolipid (about 5 mg) obtained for example by preparative chromatography[65], is heated in a screw-closing test tube with 2 ml of 5 % solution of hydrochloric acid in methanol. After cooling, the solvent is evaporated and the residue is dried in a dessicator over sodium hydroxide and dissolved in chloroform-methanol (1:2) to yield a 2–3 % solution. An amount of 1 µl is applied to Chromarods S II and developed in a system of chloroform-methanol-15 % ammonium hydroxide system (84.5:14.1:1.4) to a height of 6 cm and then in n-hexane-diethyl ether (71:29) to 10 cm. Methylated saccharides (glucose, galactose, R_F 0.05), sphingoid bases (sphinganine, sphingenine, R_F 0.38), the methyl esters of hydroxyacids (e. g. methyl-12-hydroxystearate, R_F 0.78) and the methyl esters of the fatty acids (0.82) are separated from the start. This method is also useful for the determination of the molar ratio between phosphorylcholine (R_F 0.00), sphingenine (0.51) and the methyl ester of the fatty acids (0.82) in methylated sphingomyelin, as well as for identification of other phospholipids, e. g. phosphatidylcholine. The methylated mixture is then gradually separated in systems of ethyl acetate-formic acid (97:3, twice to 6 cm) and of n-hexane-diethyl ether (71:29, once to 10 cm). Phosphorylcholine remains at the start, clearly separated from glycerol (R_F 0.22) and the methyl esters of the fatty acids (R_F 0.84)[125].

It is sometimes difficult to select an internal standard for the application of TLC-FID to series analyses in clinical practice, where it is required to determine the contents of the individual lipids in plasma or in tissues without having to maintain precision in sample application. The internal standard is first dissolved in the solvent mixture to be used for the extraction. In addition to the compounds listed in Tables 16 to 20, other substances have also been tested with less success, e. g. tributylglycerol, vitamins A, D, E and K[48], undecylenic acid, pregnanediol[80], n-eicosane, n-dodecylbenzene, lithocholic acid[119] and methyl palmitate[112]. It seems probable that only cholesterol acetate[117, 119] and umbellipherylpalmitate[48] are of practical importance. In the single-step separation of simple lipids, octadecanol can be used, either natural[80] or synthetic, e. g. Alfol RD 18[61]. This standard is readily separated by system no 8 (Table 16) and appears in the centre of the chromatogram between free cholesterol and fatty acid. In addition, this elution system (hexane, diethyl ether) can be easily removed from the rod at a relatively low temperature (60–70 °C), i. e. under conditions where octadecanol is not volatile. The use of umbellipherylpalmitate is dependent on the use of a solvent of poor volatility(2-methyl-1-propanol, system no. 1, Table 18). Residues of this substance in the thin layer increase the detector noise and thus reduce the reproducibility of the measurement. A disadvantage of the use of methylpalmitate as an internal standard is the high sensitivity of its R_F value to diethyl ether and organic acids (formic, acetic) content in the elution mixture (see Table 17).

2.1.1.3 FID Reproducibility and Response

A number of papers describe in detail the accuracy and reproducibility of the determination of simple lipids and phospholipids. The accuracy is to a certain degree dependent on the determination of the correction factors, following from the different responses of the individual compounds in the FID[61, 77, 82, 99, 109, 126].

However, the response to a particular compound depends not only on its chemical composition but also on the detector regime and the speed of the rod through the FID (see Section 1.2.3).

Table 22 Relative TLC-FID Responses of Simple and Complex Lipids, Related to the Response for Cholesterol (carbon content 83.8 %). Recalculated literature data for a cholesterol response of 1.00.

Carbon content (%)	LPC 57.7	SM 67.3	PC 65.8	PE 65.7	MG 69.1	FA 74.9	TG 75.8	CE 82.6	Reference
theoretical response	0.69	0.80	0.78	0.78	0.82	0.89	0.90	0.98	
experi-	–	–	–	–	–	0.75	0.68	0.67	109
mental	–	–	–	–	–	0.70	0.55	1.30	48
response	–	–	–	–	–	–	0.63	0.86	80
	–	–	–	–	–	–	0.60	1.01	77
a	0.73	0.75	0.85	0.76	0.88	0.54	0.60	1.01	126

a additional responses: hexadecane 0.44; 1-hexadecanol 0.48; hexadecylpalmitate (wax) 0.72; ceramide 0.96; cerebroside 0.93.

It is not surprising, therefore, that the response for the individual lipids found by different authors is very different, as can be seen from the data in Table 22. The theoretical responses are also listed, corresponding to the carbon content in the lipid molecule. The results are related to unit response for cholesterol, which yields a higher FID response than that corresponding to its composition (as in GLC-FID). This is also confirmed by most of the results in the table, where the relative responses for the other lipids are lower than the theoretical value, except for the cholesterol ester.

The different values listed in Table 22 can arise from the fact that the response was found in some papers by measurement of standard mixtures[126] and in others by comparing with the analytical results obtained by classical biological methods[48, 77, 80], which can be accompanied by a systematic error. The response to chemically related substances, such as phosphatidylcholine, phosphatidyletha-nolamine and their lysoderivatives are, however, directly proportional to the carbon content and do not depend on the type of silica gel layer employed. It is apparent from Table 23 that the responses to lyso-compounds are about 15 % lower than those for the diacyl-derivatives.

Table 23 Comparison of the Experimental FID Responses for Phosphatidylcholine, Phosphatidyletha-nolamine, and Their Lyso Derivatives with the Theoretical Values, Calculated from the carbon content in the molecule. Experimental conditions given in Table 24[36].

Thin-layer	LPC	PC	LPE	PE
Chromarods S Chromarods S II	0.89 ± 0.04 0.91 ± 0.07	1.14 ± 0.03 1.10 ± 0.05	0.92 ± 0.03 0.93 ± 0.04	1.11 ± 0.03 1.11 ± 0.05
mean	0.90	1.12	0.92	1.11
theoretical response[a]	0.90	1.03	0.90	1.02

[a] Calculated from the percentage carbon content and related to LPC (57.7 % C – response 0.90).

Detailed information on the responses for a large number of various lipids and natural substances and their dependences on the amount analysed can be found in the work by Kaitaranta and Nicolaides[126]. These results are, of course, only informative. In normal quantitative analysis, correction factors derived from the FID response are used only exceptionally and the results are evaluated directly from calibration curve equations, preferably from the dependence of the ratio of the response for the component and for an internal standard on the ratio of their masses (see Section 1.2.3). An example of the application of these dependences in the analysis of serum lipids is given in Figure 24.

Problems connected with the reproducibility of the FID response are reflected to a certain extent in the reproducibility of the quantitative determination, which is usually lower than that for the gas chromatography of the individual lipids[61]. A large number of papers deal with the reproducibility of TLC-FID[37,48,61,77,80,82,114,119]; the coefficient of variation averages about 5%. Some authors give higher values[80,82,119], some lower values[61,127]. Mills[114] and Ackman[35] analysed in detail the individual factors affecting the reproducibility of these determinations.

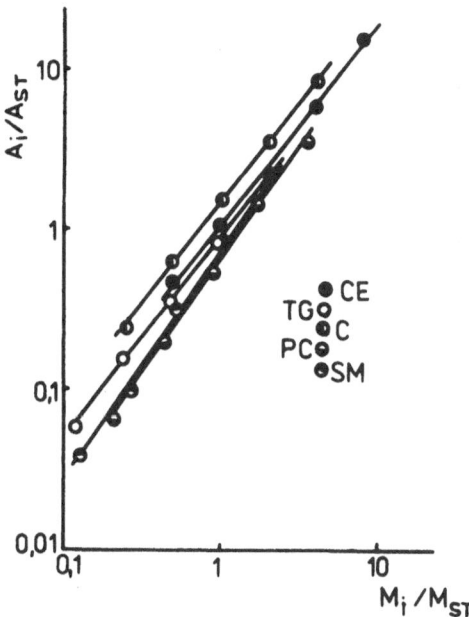

Fig. 24. Calibration straight lines for analysis of lipids using system no. 4. Table 20 (on Chromarods S II[99])

The main reason for the lower reproducibility of TLC-FID compared to GLC-FID lies in the differences in the separation abilities of the individual thin layers and in the detector design; there is no guarantee that part of the separated component will not vaporize prior to entering the FID or that all the ions produced in the hydrogen flame will be collected by the collector electrode. These problems are being studied in a number of laboratories and were discussed in detail in section 1.2.3.

The reproducibility of the determination, of course, also depends on the amount of the component applied. It is apparent from Figure 25 that the variation coefficient increases rapidly when the amount of lipid applied falls below 1 µg[61]. The determination of smaller amounts can show poor reproducibility. When large amounts are applied, the separation capacity of the layer may be exceeded, with incomplete combustion of some of the components in the FID.

The dependence of the reproducibility on the amount applied, given in Figure 25, was constructed from the results of the analysis of cholesterol ester with fatty acids, free cholesterol, triacylglycerol, free fatty acids and the sum of the phospholipids on Chromarods S II using elution system no. 4 (Table 16). As the determined coefficients of variation were not markedly dependent on the type of lipid, the relationship is characterized by a single curve.

The reproducibility is also dependent to a certain degree on the type of layer used and on the properties of the given rod production batch. Coefficients of variation for the determination of the peak areas of phospholipids determined by TLC-FID on silica gel layers (especially on Chromarods S) are usually lower than on alumina (Chromarods A). In the latter case the separated zones are broader, the peaks are not symmetrical and the noise level is higher. The reproducibility of the determination of the R_F values is, however, quite good and comparable with other types of rod (Table 24).

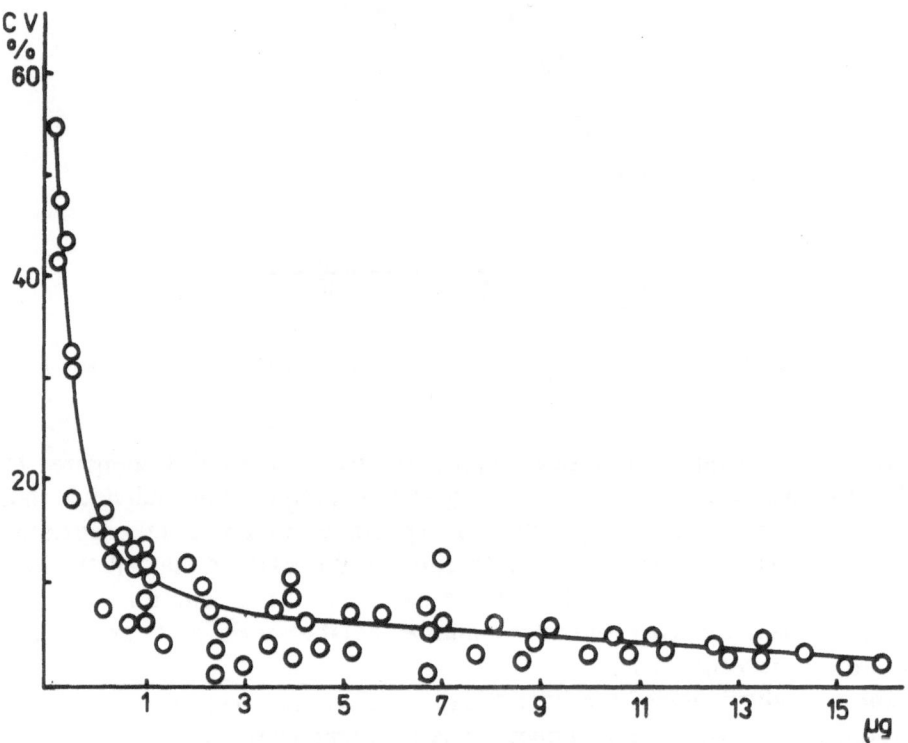

Fig. 25. Dependence of the reproducibility of the TLC-FID of lipids in blood plasma on the amount of sample applied[61]

Table 24 Reproducibility of the Determination of Phospholipids on Chromarods S and S II and A. System no. 8, Table 19, Separation of the individual phospholipids expressed as the retention times at a Scanning Speed of 0.42 cm s^{-1}. Results evaluated by integrator HP 3390A (Chromarods A)[118] or SP System 1 (Chromarods S and S II)[36]. Retention times: origin 0.0, front 56.0×10^{-2} min. An amount of 2 μl of solution applied to Chromarods A contains 1 ± 5 mg of each lipid per ml; amount of 1 μl of standard mixture contains 1.9 mg LPC, 1.8 mg PC, 2.3 mg LPE and 1 mg PE per 1 ml solution applied to Chromarods S and S II.

(a) Reproducibility of the retention times and peak areas for phospholipids on Chromarods A[118].

Rod no.	Retention time (min $\times 10^{-2}$)				Relative area $\times 10^{-3}$			
	LPC	PC	LPE	PE	LPC	PC	LPE	PE
1	50	43	35	21	32.7	34.0	9.6	23.0
2	52	44	34	21	35.5	30.7	9.7	22.6
3	52	42	34	21	33.7	29.7	10.7	24.3
4	50	41	32	19	35.4	28.3	11.5	24.8
5	51	42	34	21	33.5	32.1	12.2	22.2
6	51	43	34	20	31.2	31.3	10.3	27.1
7	51	42	33	19	30.5	27.5	9.5	29.3
8	50	42	33	19	32.7	32.1	12.0	23.3
9	52	44	35	19	31.4	33.8	12.4	22.3
10	52	43	33	19	27.4	31.1	10.0	28.9
mean	51.1	42.6	33.7	19.9	32.4	31.1	10.8	24.8
S D	0.87	0.96	0.90	1.0	2.4	2.1	1.1	2.7
C V %	1.7	2.2	2.7	5.0	7.5	6.8	10.1	10.9

(b) Reproducibility on Chromarods S[36]

Rod no.	Retention time (min $\times 10^{-2}$)				Relative area $\times 10^{-3}$			
	LPC	PC	LPE	PE	LPC	PC	LPE	PE
1	44	37	30	23	19.6	26.4	27.3	26.7
2	43	37	30	23	19.8	26.9	27.0	26.3
3	45	38	31	23	20.5	26.3	26.8	26.4
4	44	38	30	22	21.7	26.7	26.6	24.9
5	44	38	31	22	21.3	26.2	26.1	26.4
6	43	37	30	22	22.7	26.0	25.4	25.9
7	43	36	29	22	22.0	25.5	27.1	25.4
8	44	37	29	23	22.0	25.4	27.7	24.9
9	44	37	29	22	22.3	24.7	27.2	25.8
mean	43.7	37.2	29.8	22.4	21.3	26.0	26.8	25.9
S D	0.63	0.62	0.73	0.49	1.02	0.73	0.76	0.59
C V	1.4	1.7	2.4	2.2	4,8	2.8	2.8	2.3

(c) Reproducibility on Chromarods S II[36]

Rod no.	Retention time (min × 10^{-2})				Relative area × 10^{-3}			
	LPC	PC	LPE	PE	LPC	PC	LPE	PE
1	43	36	31	24	21.1	26.5	27.6	24.8
2	45	38	31	24	21.6	24.0	26.7	27.7
3	45	38	30	23	21.6	23.0	26.9	28.5
4	44	38	31	23	23.3	27.0	25.4	24.3
5	45	38	31	24	25.3	23.2	26.8	24.7
6	44	38	31	25	19.9	24.1	28.9	27.1
7	44	38	31	25	22.1	25.6	26.9	25.4
8	44	38	31	27	21.6	26.0	27.5	24.9
9	45	38	32	22	20.8	26.9	26.6	25.7
mean	44.3	37.8	31.0	24.8	21.9	25.1	27.0	26.0
S D	0.67	0.62	0.47	1.41	1.78	1.25	1.16	1.38
C V %	1.5	1.7	1.5	5.7	8.1	5.0	4.3	5.3

The values in Tables 24 (b) and 24 (c) are taken as a basis for calculation of the FID responses (see Table 23).

2.1.1.4 Applications

As mentioned in the introduction to this section, TLC-FID has found broad application in the analysis of lipids in medicine and biological research. This is especially true in the research on the compositions of blood plasma lipids related to irregularities in lipid metabolism in the liver. Extraction of lipids from plasma is simple, as is the composition of the actual lipids. In addition to free and bound cholesterol and triacylglycerol, the main component is phosphatidylcholine and sometimes also sphingomyelin; the contents of the other phospholipids (phosphatidylethanolamine, lysophosphatidylcholine, phosphatidylserine, phosphatidylinositol) are practically negligible. In addition to the contents of free and bound cholesterol, an important factor in the study of hyperlipoproteinemia is the ratio of sphingomyelin to phosphatidylcholine[128]. This ratio is considered to be one of the important criteria in classification of patients according to the Frederickson system[129]. In contrast to the very tedious original methods (biochemical, GLC, TLC), the TLC-FID method can be used to obtain information much faster. The lipid contents can be found from a single chromatogram using a single standard. We have employed this method in the long-term study of the role of soya lecithin and phosphatidic acid salts in preventing and study of regression of various types of dyslipoproteinemia[130]. Further, recently studies have been undertaken of the effect of cheese containing phospholipids on changes in lipid metabolism in patients suffering from chronic and acute hepatitis[131]. Some interesting results are listed in Table 25 and Figure 26.

Table 25 Plasma Lipids and Their Ratios for Normal Patients (N) and Those with Hyperlipoproteinemia (HPLP). The results in the first four rows were obtained from sets of 14 and 32 patients by classical methods[128], in the rest by the TLC-FID method (Chromarods S II, System no. 3, Table 20) from sets of 5—10 patients, 23—60 years old[130].

Type HPLP	Reference	Lipids (mg per 100 ml blood)				
		CT	TG	PL	TG/CT	SM/PC
N		190 ± 5	86 ± 6	229 ± 5	0.45 ± 0.02	0.252 ± 0.004
IIa	128	298 ± 10	101 ± 4	279 ± 9	0.35 ± 0.02	0.321 ± 0.010
IIb		299 ± 7	182 ± 8	288 ± 7	0.61 ± 0.03	0.304 ± 0.012
IV		267 ± 17	318 ± 34	304 ± 15	1.20 ± 0.10	0.206 ± 0.008

Type HPLP	ref.	C	CE	CT	TG	PL	TG/CT	SM/PC
N	130	35 ± 5	200 ± 20	158	71 ± 15	205 ± 30	0.45	0.242
IIa		88 ± 30	308 ± 10	278	96 ± 30	250 ± 30	0.34	0.327

CT total cholesterol.

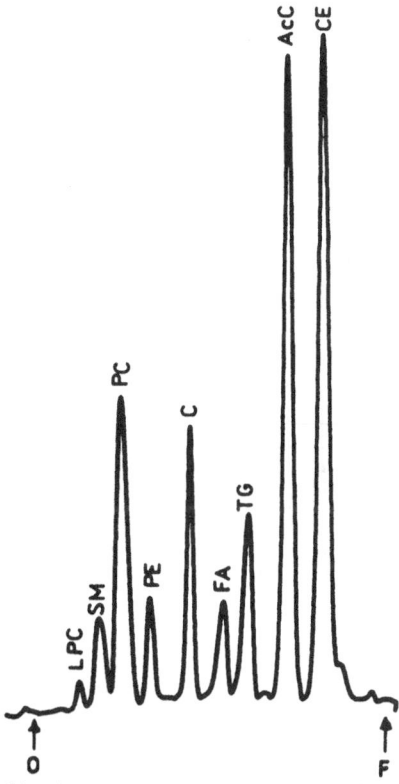

Fig. 26. TLC-FID of lipid in blood plasma of a person with primary hyperlipoproteinemia type II b. Chromarods S II, system no. 3, Table 20, developed 3(a), twice to 5 cm; 3(b), once to 6 cm; 3(c), once to 10 cm. Scanning speed 0.42 cm s^{-1}, recorder sensitivity 100 m V f. s., AcC – cholesterol acetate.

Figure 26 depicts a TLC-FID chromatogram of the blood plasma lipids of a person with primary hyperlipoproteinemia, type IIb. Quantitative evaluation of the recording yielded the following contents of the individual lipids (mg per 100 ml plasma): cholesterol ester (CE) 365, triacylglycerol (TG) 212, fatty acid 50, cholesterol (C) 55, phosphatidylethanolamine + phosphatidylinositol + + phosphatidylserine 40, phosphatidylcholine 182, sphingomyelin 55, lysophosphatidylcholine 14, total cholesterol 281, total phospholipids (PL) 291. The characteristic ratios of the important lipids (for comparison with Table 25) are: CT/PL 0.97, TG/CT 0.75, SM/PC 0.302.

TLC-FID has also been used to study the changes in the lipid profile in rat liver cell membranes resulting from the effects of organic solvents. A suspension of isolated rat liver cells was reacted with FMA (the complex of fluorescein with mercuric acetate) twice for 30 min and was centrifuged at 1500 g to yield a

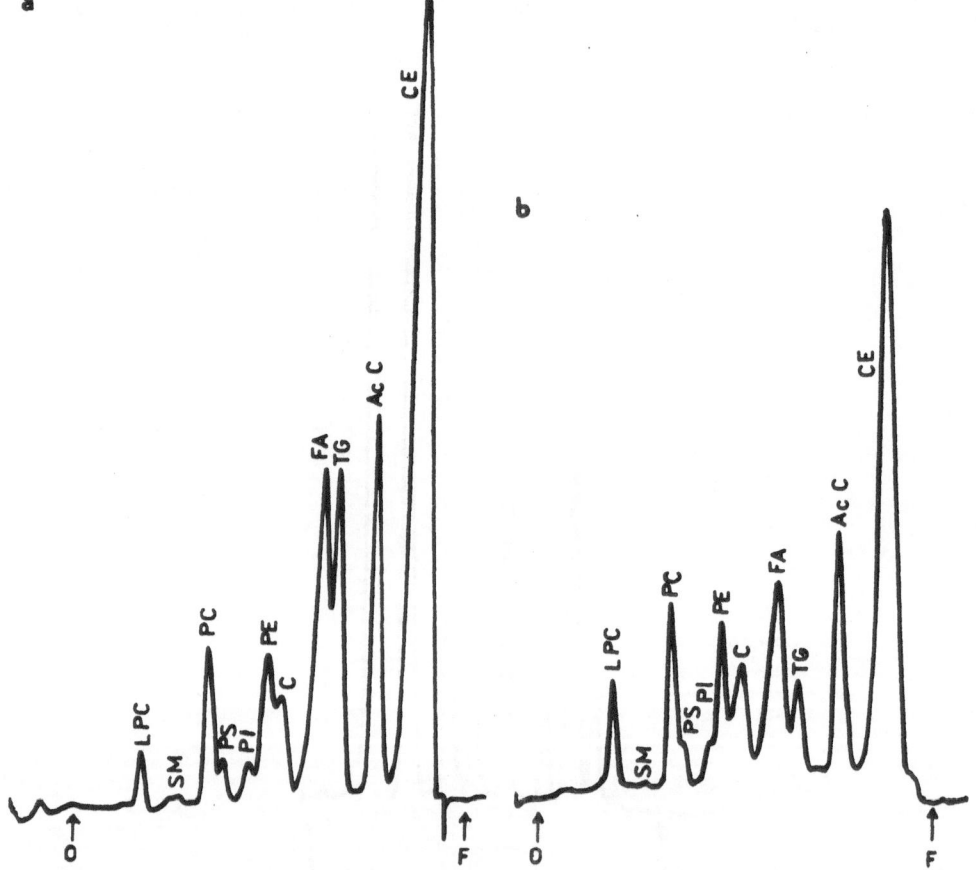

Fig. 27. TLC-FID of lipids from the membranes of rat hepatocytes. a: original hepatocytes, b: after application of carbon tetrachloride[132]. Chromarods S II, system no. 3, Table 20, elution, detection rate, recorder sensitivity as in Figure 26.

saccharose solution containing a small amount of calcium chloride. The sediment was freed of cytoplasmic components using a salt solution and was recentrifuged. The sample was prepared for analysis by extracting twice with a 20-fold excess of chloroform and methanol (2:1) and evaporated at 40°C[132]. Figure 27 depicts the TLC-FID recording of the lipids in the original and damaged hepatocytes.

Application of organic solvents led to an increase in the total amount of phospholipids and free fatty acids. The relatively high content of lysophosphatidylcholine and negligible content of sphingomyelin are interesting. The presence of a large amount of neutral lipids, which should not be present in the sample in amounts greater than 40 %[133], was explained[132] by the presence of cytoplasmatic lipids in the sample.

It has been known for some time that certain properties of the blood platelets play an important role in coronary diseases; these include their adhesive and aggregation abilities, resulting from the composition and physical state of their membranes. Periodic study of the lipid distribution, which is directly related to the membrane fluidity[134] would be very useful in the systematic study of this problem. Japanese authors[119] have stated that the TLC-FID method is useful in this connection, compared to the commonly used analytical procedures, primarily because the contents of all the lipids can be found simultaneously in a single analysis, so that their ratios can be determined with a smaller error. Chromatography in a chloroform-methanol-water system (system no. 9, Table 19) led to good separation of the individual phospholipids from neutral lipids, except for phosphatidylserine and phosphatidylinositol, which formed a critical pair. The blood platelet membranes of healthy individuals contain about 15 % (by weight) of lipids, of which 35 % is phosphatidylcholine, 25 % phosphatidylethanolamine, 19 % sphingomyelin, 20 % phosphatidylserine + phosphatidylinositol and the remainder is primarily free cholesterol[119].

Takemoto[101, 135] is mainly responsible for the application of the TH-10 Iatroscan in research on hepatobiliary disorders (acute and chronic hepatitis, cirrhosis and cancer of the liver and obstructive jaundice) and on haematological diseases (such as iron-deficiency anaemia, hereditary spherocytosis, acantocytosis, etc.). The symptoms of these diseases can also be found in the lipid composition of the erythrocyte membranes. For example, diseases of the liver, especially obstructive jaundice, are characterized by an increase in the content of free cholesterol and phosphatidylcholine, together with a corresponding decrease in the phosphatidyl ethanolamine and sphingomyelin contents. The increase in the ratio of phosphatidylcholine + sphingomyelin to phosphatidylethanolamine + phosphatidylserine is especially important in cancerous forms[136]. On the other hand, a tendency for the overall lipid content to decrease has been observed in haematological diseases; primarily, the lipids found in the inner part of the membrane decrease. These are phosphatidylethanolamine and phosphatidylserine[135]. Both these papers contain detailed tables of analytical results, including a description of the course of acute hepatitis. The author attempted to explain changes in the lipid concentrations in the external (phosphatidylcholine, sphingomyelin), central (free cholesterol) and inner (phosphatidylethanolamine, phosphatidylserine) parts of the membrane. For example, he explains the increase in the lipid in the membrane in liver diseases in terms of an increase in concentration of phosphatidylcholine and free cholesterol in the blood. The methodology of TLC-FID analysis of lipids in erythrocyte membranes, together with the isolation procedure, can be found in the original paper by Takemoto[101] and another Japanese author, Tadano[83]. A TLC-FID trace is shown in a paper by Ackman[35]. A study was carried out on the dietary effects of a high fat content on the distribution of phospholipids (phosphatidylcholine, phosphatidylethanolamine and sphingomyelin) and glycolipids (monoglucosylceramide), which were isolated from the plasmatic membrane of the intestinal mucosa, the liver and the brain by elution with system no. 4 of Table 19[115, 137]. Nonadecane was used as an internal standard[137].

The lipid composition in the lysomal membranes of the atheromatous aorta differs from that of normal biological membranes. It follows from analysis using Chromarods S II in a chloroform-methanol-water system (16:8:1) that the contents of sphingomyelin, phosphatidylcholine and free cholesterol and cholesterol esters increase in the membranes of the cells of the affected aorta[138].

Tsuchiya et al.[103] applied TLC-FID to the determination of lipids in the erythrocyte membranes of patients with mycoplasma pneumonia. They separated the extract using a system containing ammonia (chloroform-methanol-water-ammonium hydroxide, 60:40:1.7:0.1); good separation of sphingomyelin from phosphatidylcholine (R_F values of 0.12 and 0.21) and phosphatidylserine from phosphatidylethanolamine and cholesterol (R_F 0.25, 0.39 and 0.67) was obtained.

Hiramatsu and Arimori[107] determined the lipid contents in the lymphocytes of a healthy individual. A very large amount of blood was formerly used for such a determination (about 250 ml). A volume of 20 ml of blood is now sufficient for the same analysis using the Iatroscan instrument. Lymphocytes were separated by the Böyum method[139] and rinsed twice in 0.15 M sodium chloride with 1 mg EDTA per ml of solution. The resulting solution was then extracted with a chloroform-methanol (2:1) mixture together with the internal standard (cholesterol acetate, 1 mg ml^{-1}). The extract was freed of contamination by washing with a salt solution and was filtered and dried in a rotary vacuum evaporator at a temperature below 37 °C. Neutral lipids were determined on Chromarods S using an n-hexane,diethyl ether (9:1) system, and phospholipids using a chloroform-methanol-water mixture (60:30:3.5) containing 2,4-di-tert-butylphenol (50 mg ml^{-1}). A total of $2 \cdot 10^7$ lymphocytes contained 1–1.5 mg of lipids, approximately half of which were neutral lipids (TC, CE and C), the remainder being phospholipids (PC, PE, SM and small amounts of PS and PI). Phospha-tidylethanolamine should be primarily a plasmalogen structure (1-alkenyl-2-acylglycerophosphoethanolamine) which, however, has the same R_F value as the corresponding diacyl derivative under normal conditions. Plasmalogens are also an important component of the lipids of other cell membranes, such as eryth-rocytes[140], sperm and the cells of the central nervous system. Some time ago, a paper was published describing a procedure for the separation of ethanolamine plasmalogen from phosphatidylethanolamine following mild hydrolysis by the vapours of hydrochloric acid[141]. This is a TLC-FID adaptation of the original work by Horrocks, based on two-dimensional TLC of the products of the hydrolysis of samples containing plasmalogens[142]. Chromarods S onto which lipid samples from the synopsomal membrane of rat brains were applied, were first developed in a system containing light petroleum (b. p. 50–110 °C) and diethyl ether (85 : 15) in order to separate neutral lipids and nonadecane (as an internal standard). After pyrolysis in the FID by selective scanning, an un-separated zone of phospholipids remained at the start. Half the rods were then suspended in the glass frame over the surface of hydrochloric acid and exposed to the fumes in the chromatographic tank for 5 min in such a way that the phospholipid band was about 4–5 cm from the surface. The rods were then dried for 5 min at 70 °C, reactivated by scanning the upper part of the rod and eluted in system no. 4 (Table 19). After redrying (5 min, 70 °C), the entire length of the rod was scanned. The other half of the rod was analysed in the same way, but without hydrolysis in a hydrochloric acid atmosphere. Hydrolysis of the plasmalogens yielded two new peaks on the chromatogram: lysophosphatidy-lethanolamine (R_F 0.44) and a long-chain aldehyde (R_F 0.84) split from the plasmalogen. Quantitative analysis indicated that, under these conditions (5–10 min exposure to hydrochloric acid vapours), the ester-bonded fatty acids remain unchanged, i. e. the concentrations of phosphatidylcholine, sphingomye-lin and phosphatidylserine do not change.

The TLC-FID of lipids extracted from human skin is an example of successful utilization of partial scanning in the chromatography of complex mixtures. The less polar substances (squalene, cholesterol esters and waxes) are first chromato-graphed in an n-hexane-benzene system (1 : 1). After partial scanning, triacyl-glycerols, free fatty acids, free cholesterol, monoacylglycerol and a mixture of phospholipids are then separated in the next elution cycle (using an HDF system, 70 : 30 : 1). This method requires two internal standards, thymol in the first elution and anisalcohol in the second[143]. This paper also contains a table comparing the reproducibility of the determination of the above lipids using planar TLC and thin-layer chromatography with flame ionization detection. The TLC-FID results are found to be more reproducible.

Skin lipids can also be separated in a single step on one rod by stepwise elution with the above-mentioned solvent systems on Chromarods S II. The first system

114

is allowed to flow to a point 11 cm above the start and the second, more polar system moves to a distance of only 5 cm[35]. Figure 28 gives an example of such a chromatographic procedure. The major lipids in the extract are triacyl-glycerols, waxes and squalene; the minor components are cholesterol and phospholipids.

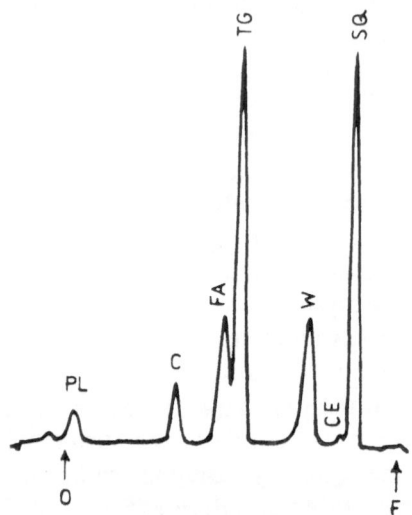

Fig. 28. Separation of skin lipids on Chromarods S II by stepwise elution with n-hexane-benzene (1:1) to 11 cm and n-hexane-diethyl ether-formic acid 70:30:1 to 5 cm. W wax, SQ squalene

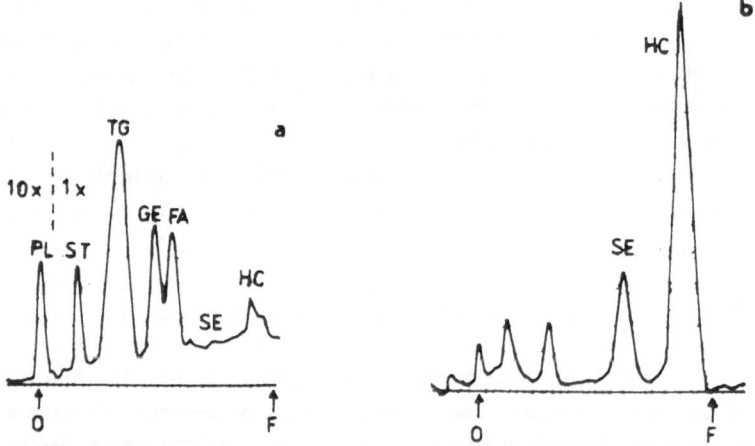

Fig. 29. TLC-FID of lipids from the crustacean Corophium volutator on Chromarods S[144]. Solvent composition given in the text: (a) chromatogram of total lipids (primarily phospholipids, see 10-fold decreased sensitivity. (b) chromatogram of extract of silica gel from the upper part of a planar layer (see text). The three small peaks at the bottom of the chromatogram correspond to triacylglycerol, sterol and phospholipid esters

Ackman et al.[35] utilized the TLC-FID technique for the analysis of very small amounts of hydrocarbons and esters of cholesterol in lipids from small crustaceans of the species Corophium volutator, which have a weight of about 0.45 g and contain only 1–1.5 % (by weight) fatty substances[144]. This is a sufficiently large amount for the determination of phospholipids and triacylglycerols, but not enough for the minor lipids (squalene, sterol esters, see Figure 29(a)). Quantitative analysis of substances present in small amounts would require application of a larger amount of sample, which would exceed the separation capacity of the Chromarods. The authors solved this problem by preliminary sample separation on a thin-layer plate using dichlorofluorescein as a detection reagent. They then scraped off the whole region of the layer lying above triacylglycerol. This layer was extracted and the concentrated extract was applied to Chromarods S and eluted stepwise first using petroleum ether-benzene-formic acid (64:14:1) and then petroleum ether-diethyl ether-formic acid (97:3:1). This procedure yielded better quantification of hydrocarbons and cholesterol ester (Figure 29(b)), and also demonstrated that, in classical chromatography, part of the phospholipids and acylglycerols migrate to the upper part of the layer, where they are not found by normal detection methods. It follows from this work that TLC-FID is more useful for control of the purity of substances (and of fractions from preparative chromatography) than planar TLC.

Systems of the n-hexane-benzene type, used in the analysis of skin and shellfish lipids, are very useful for the quantitative determination of hydrocarbons in lipids from blood plasma (either a native part of the plasma or contaminating the extract). In contrast to chromatography in systems containing-n-hexane, diethyl ether and formic acid, in which hydrocarbons form a critical pair with cholesterol ester, the separation of these substances is excellent using mixtures of the type described above (Figure 30)[36].

Hazel studied the effect of temperature on the composition of phospholipids in gill tissue from thermally acclimatized rainbow trout. The extract was applied to Chromarods S II and first separated using an HDF system (40:10:1). After partial scanning from the front to 1 cm above the start, the rods were eluted in chloroform-methanol-water (120:52.5:4.5). The best separation was obtained by equilibration of the Chromarods in a chamber with 52.5 % relative humidity[145]. The applications of the Iatroscan technique in the analysis of lipids from fish and sea organism will be discussed in further detail in the chapter on the applications of TLC-FID in the food industry (Section 2.2.1).

Kramer et al.[146] studied optimization of conditions for analysis of lipids in biological materials. It was found that simple lipids can be separated using stepwise elution in systems of HDF (95:5:1) and 1,2-dichloroethane-chloroform-formic acid (92:8:0.1). Phosphohpids were separated well using chloroform-methanol-water systems (68 5 · 29 · 2 5)

Katsu, et al. and Okutsu et al. used the Iatroscan technique to study changes

116

in the contents of the individual lipids in blood[147] and in red blood cell membranes[148] of pregnant women. They found that the lipid content in blood plasma

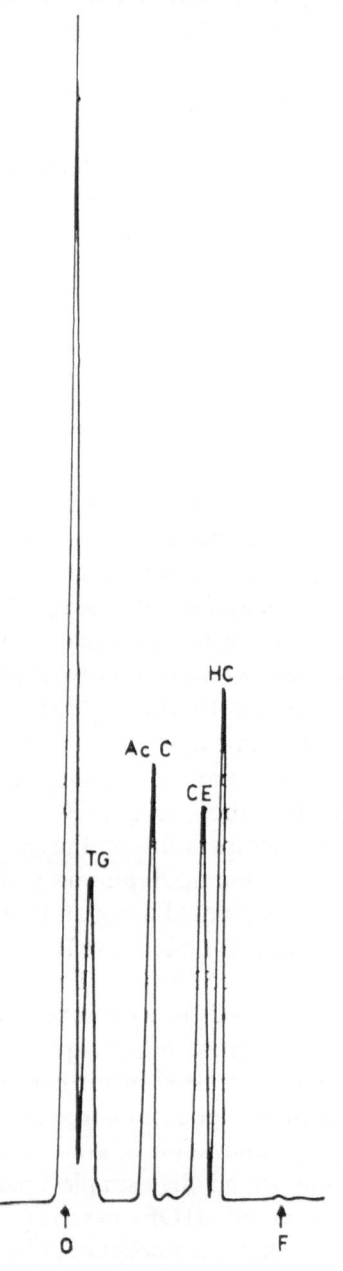

Fig. 30. Chromatogram of lipids from human plasma on Chromarods S II in an n-hexane-benzene (1 : 1) system[117]

increases throughout pregnancy and is highest prior to birth. The phospholipid content in the erythrocyte membranes also slowly increases, especially the content of phosphatidylethanolamine. This hyperlipaemia is probably a normal physiological phenomenon, independent of fat ingestion, and disappears within four weeks after giving birth. An increase in the ratio of sphingomyelin to phosphatidylcholine was observed after the eight week of pregnancy[147].

Data on the utilization of the TLC-FID method in the analysis of amniotic fluid in advanced pregnancy are numerous[106, 127, 149]. A lack of dipalmitoylphosphatidylcholine in the lung membrane can be one reason for difficulties in breathing observed in premature infants (RDS, respiratory distress syndrome). The ratio between phosphatidylcholine and sphingomyelin in the amniotic fluid, which increases during pregnancy, is one of the criteria for evaluating birth risk. It is generally assumed that the risk of RDS is minimal at phosphatidylcholine/phosphatidylserine rations of greater than 2[149]. If the mother is suffering from diabetes, then this ratio should be greater than 3. A further criterion for determining the foetal lung maturity is the phosphatidylglycerol content, which appears in the amniotic fluid after the 36th week of pregnancy and is present in minute amounts in the amniotic fluid of diabetic mothers[150, 151].

The methods used so far are based on analysis of the lipid extract from the amniotic fluid by chromatography on silica gel layers. Gluck et al.[152] suggest a procedure in which a single extraction is carried out using a mixture of chloroform and methanol (2:1). This procedure is not quantitative but, as the extractibilities of phosphatidylcholine and sphingomyelin are almost identical, their ratio remains constant[153].

The rods with the applied extract are usually eluted in a system of the chloroform-methanol-water type. In classical TLC, about one quarter of the volume of water is replaced by ammonium hydroxide, which should eliminate possible overlapping of the spots corresponding to phosphatidylcholine and sphingomyelin by those of phosphatidylserine and phosphatidylinositol[105, 154]. The separated zones are determined either by charring at increased temperatures (a silica gel layer containing ammonium sulphate) or by detection using copper acetate and heating to 180 °C[105]. This is followed by densitometric measurement of the spots or determination of phosphorus in the scraped-off zones[105, 152, 153]. This procedure is disadvantageous because of the presence of background impurities or low stability of the spot colouration, both of which adversely affect the precision of the determination. Coefficients of variation greater than 20 % are no exception[155]. Detection in analysis by the HPLC method is also problematic as phosphatidylcholine, a derivative of saturated palmitic acid, can hardly be detected at all in the ultraviolet detector[153]. Such problems are eliminated by use of the TLC-FID technique.

Two solvent systems are used to analyse the extracts; chloroform-methanol-water (60:25:3[105], 80:25:3[106] or 70:30:3[127]) to determine the PC/SM ratio and tetrahydrofuran-dimethoxymethane-methanol-ammonium hydroxide (10:3:2:1,1)[106, 127] for phosphatidylglycerol estimation. Lysophophatidylcholine can be used as an internal standard in the first elution system, dimethylphosphatidylethanolamine is the useful standard for the phosphatidylglycerol determination. As little as 2 ml of amniotic fluid suffice for the analysis[105]. The reproducibility of the analysis is quite good (the coefficient of variation for the phosphatidylcholine/sphingomyelin ratio is about 5 %)[105] and the results obtained correlate satisfactory with those from the TLC method with densitometric detection[127].

The Iatroscan TH-10 Analyzer is also useful for measuring the overall lipid content in biological and other materials. The sample solution (5 µl) is applied to the Chromarods and developed for a short time with a chloroform-methanol (1:1) mixture or in pure methanol. The lipids form a narrow zone in a short distance from the origin (focusing of the lipids). Small differences in the FID responses of the individual components has practically no effect on the signal linearity in the range 0.5–32 µg. The method is at least as sensitive and reproducible as microgravimetric measurement and permits determination of overall lipids in 10 samples in 30 min[156].

Sufficient attention has not yet been paid to the application of computer technology in the analysis of biological lipids by the TLC-FID method. The paper by Knapp and Sherill[157] is one of the few papers so far published in this field. The Iatroscan TH-10 has been interfaced with an Apple II microcomputer equipped with a noise filter, amplifier and analogue-to-digital converter. The chromatograms were displayed on a video screen and stored on floppy disks. This system has been successfully applied to the study of changes in the lipid levels from plasmatic lipiproteins as a function of dietary changes.

2.1.2 Steroid Compounds

This section deals with substances of animal and plant origin that are derivatives of cyclopentaneperhydrophenanthrene. For lucidity in the subsequent text, the structure of the basic skeleton of steroid alcohols (steroids) will be given here, with numbering of the carbons and designation of the individual rings:

TLC-FID has so far not found broad application in the analysis of steroid compounds and only simple mixtures of substances have been analysed, where differences in polarity permit simple separation by adsorption chromatography. More complex mixtures are now separated mainly using distribution chromatography by the GLC or HPLC methods.

Of the steroid alcohols (sterols), the greatest attention has been paid to cholesterol and its fatty acid esters. These studies were mentioned in the previous section.

The products of phosphorylation of cholesterol (or other steroid alcohols) can be separated on Chromarods S II by stepwise elution using an ethyl acetate-acetone-water system (78 : 19 : 3) to a height of 5 cm and HDF (96 : 3 : 1) to 10 cm to yield cholesterol phosphate (R_F 0.10), bis(cholesterol)phosphate (0.30) and unreacted initial sterol (0.55)[36]. These substances have received considerable attention in pharmaceutical research in recent years as components of liposome membranes and are more polar than the initial cholesterol[158].

Cholesterol quite readily forms esters with other acidic compounds. One of these is nitrogen dioxide, a fairly common industrial pollutant. When air containing nitrogen dioxide is breathed in, part of the cholesterol in the alveolar membranes reacts to form cholesterol nitrite, unfavourably affecting the membrane permeability. These changes can lead to serious lung damage (lung emphysema[159]). The reaction of nitrogen dioxide with cholesterol can be readily studied using Chromarods S. Elution with pure n-hexane separates the unreacted free cholesterol (R_F 0.05) from cholesterol nitrite, which migrates close to the solvent front (R_F 0.75). The method can also identify cholesterol nitrate (R_F 0.62) if formed[160].

The determination of a large number of individual sterols in natural materials, especially plants (e. g. β-sitosterol, campesterol, brassicasterol, stigmasterol, etc.) using TLC is not possible on either silica gel or alumina layers. The adsorption behaviour of these structurally very similar substances is determined primarily by the dominant polar hydroxyl in the 3-position of the steroid skeleton and differences in the length and saturation of substituents in the 17-position have only a slight effect. In this connection, gas chromatography[161] or reversed-phase TLC[162] are useful techniques. On the other hand, groups of sterols differing in the number of methyl groups in the 4-position (i. e. 4-demethylsterols, 4-methylsterols and 4,4-dimethylsterols) can readily be separated by the TLC-FID method, which is often far more useful for fast quantitative analysis than GLC. These applications will be discussed in the section on lipophilic vitamins in connection with obtaining tocopherol from plant materials.

TLC-FID has been shown to be useful in the determination of steroid bile acids, either free or conjugated with glycol or taurine[163—167].

Bile acids are the principal metabolite of cholesterol and play an important role in the formation of bile, in lypolysis and absorption of fats and fat-soluble vitamins from the intestinal lumen. The distribution of these acids in the bile is one of the criteria for evaluation of gall bladder disorders in connection with increased incidence of gall stones and cancer of the gall bladder. In biological fluids they are mostly conjugated with glycine or taurine.

Like most steroids, bile acids are hydrophobic substances and thus can be extracted from aqueous media using organic solvents such as n-butanol, a mixture of glacial acetic acid and toluene or chloroform and methanol. Other isolation procedures can also be used, such as bonding to uncharged resins, anion-exchange extraction, etc. Detailed procedures for isolation and analytical procedures are given in a number of reviews[168-170].

At present, the most effective method is considered to be gas chromatography combined with mass spectrometry (GLC-MS). However, this combination, like HPLC-MS, is rather expensive, complex and time-consuming, and unsuitable for routine clinical practice. In contrast, TLC, either in the classical planar form[170-172] or with FID is a fast technique suitable for individual and series analyses.

Table 26 lists suitable solvent systems and the separation of bile acids on Chromarods S.

In the analysis, 1–5 ml of bile is first lyophilized and the dry matter is dissolved in 1 ml of water. The solution is made acid with a drop of HCl and the bile acids are extracted into 2–3 ml of butanol. An amount of 1–2 μl of extract is applied to Chromarods S or S II[164].

Beke et al.[167] used TLC-FID in the analysis of bile acids in the duodenal contents from patients with disorders in the gastrointestinal tract, liver, bile duct and pancreas. The fluid, obtained by aspiration, was diluted with 9 M sodium hydroxide and applied to Amberlite XAD-2. Excess hydroxide was removed by washing with water and the salts of the bile acids and their conjugates were eluted with methanol (5 ml). The concentrated methanol solution with a concentration of $0.04\,mol\,\mu l^{-1}$ was applied to Chromarods S II and developed step-wise in solvent systems nos. 2 and 4 (Table 26). They found that, in healthy patients, the total amount of bile acids in the duodenal contents contains only 2 mol % free acids and that the remainder is conjugated with glycine (about 66 mol %) and with taurine (32 mol %). The molar ratio between the derivatives

Table 26 TLC–FID of Bile Acids

(a) Composition of solvent systems (vol. %)

No.	Toluene	Acetid acid	Water	Methanol	Chloroform	Reference
1	55.5	39.0	5.5	–	–	173
2	50	45	5	–	–	167
3[a]	–	25	–	–	–	165
4	–	10	5	20	65	164, 167
5	47.5	47.5	5.0	–	–	164

a   plus 50 parts (per volume) isooctane and 25 parts isopropyl ether.

(b) R_F values

Acid (or salt)	Abbrev.	R_F in system no.					
		1	2	3	4	5	2 + 4[a]
cholic	CA	0.35	–	0.23	–	0.46	0.53
deoxycholic	DC	0.57	–	0.52	–	0.62	0.70
chenodeoxycholic	CD	0.49	–	0.46	–	0.62	0.70
ursodeoxycholic	UDCA	–	–	0.40	–	–	–
glycocholic	GCA	0.09	–	–	0.53	0.15	0.41
glycodeoxycholic	GDC	0.23	0.33	–	0.65	0.34	0.47
glycochenodeoxycholic	GCD	0.15	0.15	–	0.65	0.34	0.47
glycoursodeoxycholic	GUD	–	–	–	0.65	0.27	–
lithocholic	LC	0.70	–	0.76	–	–	0.81
glycolithocholic	GLC	0.35	0.58	–	0.77	0.51	–
taurocholic	TCA	–	–	–	0.13	–	0.21
taurodeoxycholic	TDC	0.10	–	–	0.26	–	0.30
taurochenodeoxycholic	TCD	0.05	–	–	0.26	–	0.30
tauroursodeoxycholic	TUD	–	–	–	0.26	–	–
taurolithocholic	TLC	–	–	–	0.36	–	–

a   double development at 4 °C, system 2 for one hour and system 4 for 10 min[167]. Systems 1–5, Chromarods S; systems 2 and 4, Chromarods S II.

of glycine and taurine was almost 1 : 1 in the samples obtained from diseased patients.

As there is a large number of bile acids in bile (about 15), a single elution cannot yield complete separation of all the components. Particular problems are involved in connection with the conjugates with taurine, which are quite hydrophilic. It would then probably be useful to develop the rods using systems with increasing polarity, combined with partial scanning between individual elutions.

The individual and conjugate bile acids are separated well in systems containing toluene, acetic acid and water (e. g. no. 1 in Table 26) depending upon the number of hydroxyls and their positions on the steroid skeleton. The mobility clearly decreases with an increasing number of hydroxyl groups (LC > DC > CA), (see list of abbreviation in Table 26). Of the desoxycholic acids, derivatives with the second hydroxyl in the 12-position have higher R_F values (DC > CD). As expected, the conjugated bile acids are more mobile than the free acids (DC > GDC > TDC) and are separated better in a system containing chloroform, methanol, acetic acid and water (no. 4, Table 26), depending upon the number of hydroxyls and type of bonded amino acid, but independent of the position of the hydroxyl group (GLC > CDC > GCA, GLC > TLC, GDC = GCD = GUD > TDC = TCD = TUD).

The separation of bile acids from bile lipids is relatively simple. For example, an n-hexane-diethyl ether-formic acid system (66.8 : 16.6 : 16.6) can be used to separate monoacylglycerol (R_F 0.25), diacylglycerol (0.45), free cholesterol (0.53), fatty acid (0.63) and triacylglycerol (0.71) from a mixture of bile acids and phospholipids (0.16). Bile acids can be separated from phospholipids using a chloroform-methanol-acetic acid-water system (51.5 : 36.1 : 8.2 : 4.2). Bile acids (R_F 0.66) are more mobile in this system than lysophosphatidylcholine (0.22) and phosphatidylcholine (0.38)[173].

In a newer modification of this method, the sample of bile is homogenized and dissolved in 10 parts of chloroform-methanol (1 : 1). The solution is freed of bile acids by centrifuging at about 2 000 g. The sample is applied to Chromarods S and focused 1 cm from the start using the same solvent mixture. It is then developed stepwise using a chloroform-petroleum ether-methanol-acetone system (6 : 2 : 1 : 1) to a position 8–10 cm from the start (30 min) and then with acetone-water (1 : 1) to a height of 5 cm (15 min). Each system can be used up to three times. The contents of the individual components are found from calibration curves, which are linear in the range 0.25–8 μg of the individual components. A standard mixture is prepared from the same weights of the sodium salts of the bile acids (45 % by weight GCD, 18 % TCA, 27 % TDC and 10 % GCA, egg lecithin and cholesterol) and is dissolved in chloroform-methanol (1 : 1) prior to application. The relative mass responses calculated from the calibration curves are[174]: cholesterol 1.00, phospholipids 0.80, bile acids 0.51.

The reproducibility of the determination of the fatty acids, similar to that for blood plasma lipids, depends on the amount of sample applied. The coefficient of variation is about 5 % for a concentration of 0.8–2 μg of the individual bile acids and is about 2 % for higher concentrations. The amount of sample mixture applied should not be greater than 10–20 μg; otherwise, separation is incomplete. The sensitivity limit is 100 ng[167].

The number of hydroxyls and double bonds is also decisive in the chromatography of steroid hormones and cardiotonic steroids. The position of the hydroxyl group also plays a role. For example, 11 α-hydroxyprogesterone is more polar than the 17α-isomer; gitoxigenin with hydroxyls in positions 3, 14 and 16 migrates more slowly than digoxigenin with hydroxyls in positions 3, 12 and 14 (Table 27).

Table 27 TLC–FID of Sex Hormones, Hormones of the Suprarenal Gland and Cardiotonic Steroids
(a) Elution systems for the TLC of sex hormones (vol. %)

Component	System no.			
	1	2	3	4
acetone	3.8	20	–	–
benzene	96.2	–	90	–
chloroform		80	–	98
methanol	–	–	10	2
reference	175	176	176	177

(b) Elution systems for TLC of hormones from the suprarenal gland and of cardiotonic steroids

Component	System no.					
	5	6	7	8	9	10
acetone	–	–	25	–	–	–
chloroform	98.9	95	–	89.1	90	92.6
ethyl acetate	–	–	75	–	–	–
water	–	–	–	1	–	–
methanol	1.1	5	–	9.9	10	7.4
reference	176	176	175	175	175, 176	175

(c) R_F values of sex hormones

Hormone	System no.			
	1	2	3	4
etiocholanolone	0.21[a]	–	–	–
dehydroepiandrosterone	0.32[a]	–	–	–
androsterone	0.43[a]	–	–	–
testosterone	–	0.20	–	–
4-androstane, 3,17-dione	–	0.66	–	–
progesterone	–	0.67	–	–
oestriol	–	–	0.15	–
oestradiol	–	–	0.50	–
oestrone	–	–	0.68	–
11α–hydroxyprogesterone	–	0.15	–	–
17α–hydroxyprogesterone	–	0.35	–	–
17α–methyltestosterone	–	0.50	–	–
dehydroisoandrosterone	–	0.67	–	–
pregnandiol	–	–	–	0.42
(cholesterol)	–	–	–	0.78

a Chromarods A, double development.

(d) R_F values of suprarenal hormones

Hormone	System no.			
	5	6	9	10
prednisolone	0.15	–	–	–
prednisone	0.51	–	–	–
β-methasone	0.87	–	–	–
hydrocortisone	–	0.12	–	0.49
cortisone	–	0.51	–	–
desoxycortisone	–	0.87	–	–
tetrahydrocortisone	–	–	0.55	0.27
β-cortolan	–	–	0.36	0.10
tetrahydrocortisol	–	–	0.44	–

(e) R_F values of cardiotonic steroids

Steroid	System no.		
	7	8	9
gitoxigenin	0.33[a]	—	—
digoxigenin	0.47[a]	—	—
digitoxigenin	0.68[a]	—	—
digoxin	—	0.36	0.18
gitoxin	—	0.50	0.50
digitoxin	—	0.64	0.80

a Chromarods A.

Bonding of a sugar on the hydroxyl in the 3-position on the steroid skeleton reverses the migration of the two coronary glycosides, i.e. the derivative of gitoxigenin, gitoxin, has a higher R_F value than digoxin even though the composition of the saccharide chain is identical in both toxines.

In clinical practice, the use of TLC-FID in the series analysis of the contents of pregnane-diol (5β-pregnane-3α, 20α-diol) in urine is especially interesting[177].

The content of this metabolite of progesterone is much higher than normal in the urine of pregnant women (Figure 31). The method used is a modification of the classical method published in 1962 by Waldi[178].

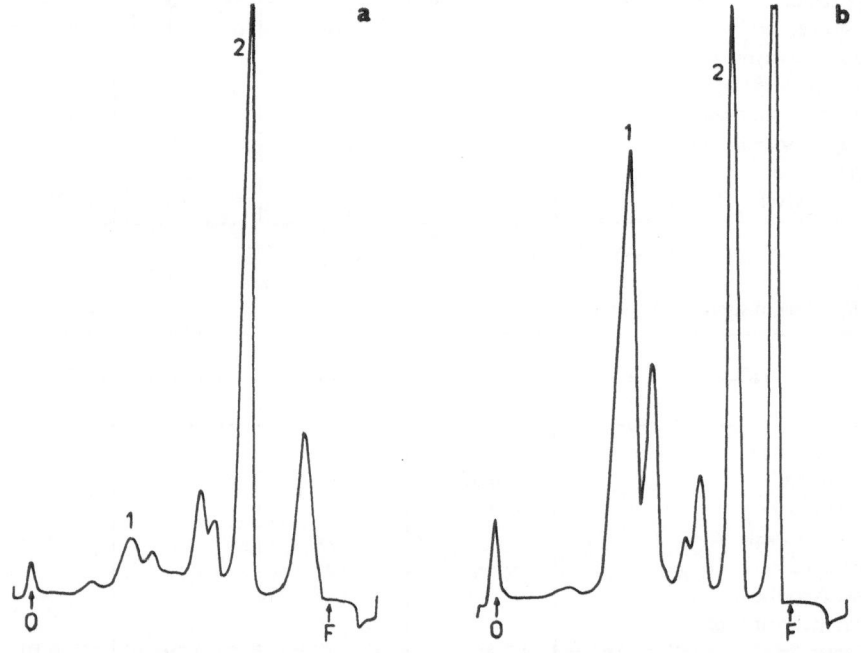

Fig. 31. TLC-FID of the metabolites of progesterone in female urine: a) non-pregnant woman, b) pregnant woman. (1) pregnanediol; (2) free cholesterol. Chromarods S II, system no. 4, Table 27

As pregnane-diol is present in urine as the ester with glucuronic acid, the ester bond must be hydrolysed prior to extraction, either chemically or enzymatically[66]. The following procedure is useful:
50 ml urine, 20 ml toluene and 5 ml concentrated hydrochloric acid are heated to 100 °C for 10 min. The solution is cooled, the aqueous phase separated and the toluene layer washed twice with 20 ml of 5 % sodium hydroxide solution containing a small amount of sodium chloride. The solution is then washed three times with 20 ml water and dried by adding anhydrous sodium sulphate. The filtrate is then mixed with the internal standard (50 µg cholesterol) and the solution is evaporated at decreased pressure. The residue is dissolved to yield a 1–2 % solution in cyclohexane.

Of the steroid saponins, primarily extracts of the ginseng root have been studied by Japanese authors. Chromatography of the extract in a methanol-toluene (8:92) system yielded chromatograms with nine sharply separated peaks of unidentified steroids; a chloroform-methanol-water (59:32:9) system (lower phase) separated the saponins[179, 180].

2.1.3 Vitamins

The literature contains almost no information on the use of TLC-FID for the determination of the contents of hydrophilic and lipophilic vitamins in natural substances, pharmaceuticals and foodstuffs. This is probably because of the low stability of most vitamins and problems connected with the reproducibility of the response in the flame ionization detector. Published papers indicate the possibility of employing TLC-FID in this area but are limited to the analysis of

Table 28 TLC–FID of Hydrophilic Vitamins (R_F Values)

	System	Vol. %	Chro- ma- rods	Vitamin							Refe- rence
				C	B_1	B_2	B_{5K}[a]	B_{5A}[b]	B_6	B_{12}	
1	benzene acetone methanol ammonium hydroxide	58 19 19 4	S	0.00	0.11	–	–	0.52	0.23		182
2	water ammonium hydroxide	99.9 0.1	S II	0.90	0.10	–	–	–	0.72	–	180
3	water ammonium hydroxide	99.95 0.05	S	0.76	0.02	–	–	0.58	–	0.40	175
4	acetone water	90 10	S	0.65	0.33	0.18	0.49	0.80	0.95	0.05	176

a nicotinic acid; b the amide of nicotinic acid.

model mixtures of pure vitamins. Most measurements have been carried out by the Iatron company[175, 176, 180—182].

Hydrophilic vitamins can be analysed either using more complicated polar systems (no. 1, Table 28) or simple elution agents, consisting of water with addition of small amounts of ammonia (nos. 2 and 3, Table 28). The polarity of the system has a great effect on the position and order of the individual vitamins on the chromatogram. For example, in system no. 1, the zone of ascorbic acid remains at the start, while when the elution is carried out using a dilute aqueous solution of ammonia it appears at the front.

Aqueous acetone has very good separating ability (system no. 4, Table 28). Practically all hydrophilic vitamins are separated along the whole length of the chromatographic rod.

We have not obtained very good results in the TLC-FID analysis of hydrophilic vitamins. The separation was insufficient using the systems listed in Table 28; the individual peaks are rather broad and irregularly shaped.

Even less data is available on the chromatography of lipophilic vitamins. Here the published data is limited to examples of separation of model mixtures with quite simple eluents that were found to be useful in classical TLC on silica gel layers[175, 181]. For example, pure chloroform can be used for good separation of α-tocopherol from its derivative, acylated nicotinic acid (system no. 2, Table 29) A solution of n-hexane in benzene (no. 3, Table 29) separates a mixture of tocol and tocopherols depending upon the number of methyl groups on the benzpyrane skeleton. The trimethyl derivative of tocol is adsorbed least and the original tocol most. The positional isomers of dimethyltocol, i.e. β-tocopherol (5,7-dimethyl) and γ-tocopherol (7,8-dimethyl) form a critical pair on both silica gel (Chromarods S II) and alumina (Chromarods A). In addition, the given solvent system cannot even be used to differentiate between tocol and its monomethyl derivative δ-tocopherol, on alumina layers.

Condensates, obtained by deodorizing edible vegetable oils by hot steam, are a technically interesting source of tocopherol and other biologically active

Table 29 TLC-FID of Lipophilic Vitamins (R_f Values)

No.	System	Vol. %	Chro-ma-rods	Vitamin							Refe-rence
				A	D_2	E_α^a	E_β	E^a	E_δ^a	T^b	
1	benzene chloroform acetone	88.5 8.8 2.7	S	0.34	0.44	0.62	—	—	—	—	181
2	chloroform[c]	100	S II	—	—	0.77	—	—	—	—	175
3	benzene[d] n-hexane	96 4	S II A	— —	— —	0.44 0.50	0.38 0.38	0.38 0.38	0.27 0.24	0.18 0.24	175

a α to δ-tokopherols; b tocol; c plus the ester of α-tocopherol with nicotinic acid (R_f 0.45); d plus α-tocopherol acetate (R_f 0.50).

substances. A very complex mixture is obtained, containing tocopherol as well as dozens of various compounds including hydrocarbons, 4-demethylsterol, 4-methylsterol, triterpenic alcohols (i. e. 4,4-dimethylsterols), sterol esters with fatty acids, etc.[161]. The condensate can even contain derivatives of furan, thiophene and phenols and phenolic acids[183]. Most of these groups of substances can be quite readily separated into their individual components using capillary gas chromatography of the silanized fractions, obtained by preliminary preparative chromatography of the samples on silica gel G layers using an n-hexane-diethyl ether system $(3:2)$[161]. It is, however, usually preferable, if quantitative data are available on the contents of entire groups in the sample. This information is yielded by the TLC-FID method (Figure 32)[184].

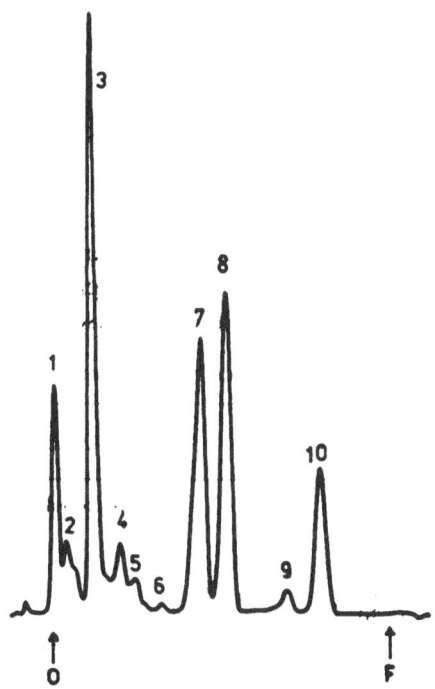

Fig. 32. Example of the application of TLC-FID to the analysis of deodorant concentrates. Chromarods S II, system no. 3, Table 29. Scanning speed 0.31 cm s[-1]. Peak identification in the text

An n-hexane-benzene system was used to separate the group of 4-demethyl-sterols (no. 3, R_F 0.14) from triterpenic alcohols (no. 4, R_F 0.22) and from a mixture of triacylglycerols and fatty acids (no. 5, R_F 0.26). In addition, tocopherols (no. 6, δ-tocopherol, R_F 0.34, no. 7, a mixture of β-and γ-tocopherols, R_F 0.44, and no. 8, α-tocopherol, R_F 0.53), the esters of sterols with fatty acids (no. 9, R_F 0.66) and a mixture of hydrocarbons (no. 10, R_F 0.75) were also separated. Methylsterols (e. g. citrostadienol) lie at the foot of the peak of 4-demethylsterols (R_F ca. 0.16). Peaks 1 and 2 correspond to more polar substances than sterols (e. g. polyglycerols, monoacylglycerols) and have not yet been identified.

Fatty acids can be shifted from the zone of triacylglycerols by preliminary methylation of the sample using diazomethane; the methyl esters formed then appear as a separate peak close to α-tocopherol (R_F 0.58)[184].

2.1.4 Alkaloids and Purine Bases

Alkaloids, products of plant metabolism, are important components of a large number of pharmaceutical products. They are mostly organic nitrogen-containing bases with complex structure, containing one or more heterocyclic rings per molecule. Polar groups present include free or esterified carboxyl, hydroxyl, keto, amino and other groups. In acid medium they form salts that are soluble in water and lower alcohols; in basic medium they form bases that are soluble in acetone, ethyl acetate, chloroform and diethyl ether. The properties of alkaloids are similar to those of purine bases which are widely distributed in plant and animal matter as components of nucleosides and nucleotides.

Alkaloids can usually be separated from purine bases using distribution chromatography on paper and silica gel impregnated with formamide or a-mines; they are chromatographed as the free bases on these materials.

Table 30 TLC–FID of Some Alkaloids and Purine Bases

No.	System	Vol. %	Chroma-rods		Alkaloids and Bases				Refe-rence
1	ethyl acetate dimethylamine N,N-dimethyl-formamide ethanol	40 20 20 20	S	R_F	tropincopolaminhydrobromide 0.59		0.72		181
2	methanol	100	S II	R_F	caffeine 0.56	salicylic acid	0.78		180
3	chloroform dimethylamine	96.8 3.2	S	R_F 0.90 0.78 0.67 0.56	aconitine thebaine reserpine caffeine	R_F 0.43 0.31 0.20 0.09	strychnine brucine codeine quinine		176
4	ethyl acetate acetone ammonium hydroxide	65 33 2	S II	R_F	quininehyd-rochloride 0.42		quinidine sulphate 0.50		185

So far, little data has been published on the TLC-FID analysis of alkaloids and related substances. Like chromatography on classical thin layers, separation of basic alkaloids on Chromarods S and S II is carried out using polar solvent systems containing dimethylamine (Table 30)[176, 180, 181].

Dimethylamine must be removed from the rod prior to detection by second elution in a non-polar solvent, e. g. n-pentane[176].

The determination of purine bases, especially caffeine, in mixtures with analgesics and sedatives will be discussed in the subsequent text.

2.1.5 Amino Acids

A few years ago, Japanese authors considered using TLC-FID in clinical practice for the determination of amino acids in some biological materials[176, 186]. However, only some amino acids could be separated, using n-propanol-water or n-propanol-ammonium hydroxide systems (Table 31). The separation is complicated by the rather broad zones obtained, especially for valine and β-alanine.

Table 31 The TLC–FID of Aminoacids (R_F Values) The meaning of the abbreviations is given in the List of Symbols.

No.	System	Vol. %	Chro- ma- rod	Tyr	Phe	Arg	His	Try	Val	Ser	Ala	Met	Refe- ren- ce
1	water n-propanol	20 80	S	0.08	0.26	0.48	0.68	0.88	0.76	0.48	0.16	0.50	176
2	ammonium hydroxide n-propanol	33 67	S	0.54	0.68	0.30	0.59	0.66	0.64	0.49	0.60	0.71	186

Relatively large problems are involved in the quantitative determination of the individual components in the FID. It follows from the papers available[186] that aminoacids behave in the flame ionization detector as low molecular weight substances with a non-linear relationship between the quantity and the response, leading to a considerable decrease in the reproducibility of the determination. Thus it has been recommended that the amount of each amino acid in the chromatographed mixture be found using a calibration curve depicting the dependence of the ratio of the sample peak area to that of an internal standard (e. g. histidine) on the ratio of the mass of the analysed substance to that of the internal standard. It is apparent that the concentration of the internal standard in the sample solution must be maintained constant. Otherwise, the calculation becomes inaccurate unless the internal standard has identical non-linear characteristics to the analysed amino acids. It is rather difficult to maintain these conditions in practice. A solution could be to increase the molecular weight of the amino acids by conversion to less polar derivatives. As in GLC[187], this would also probably improve the separation of some components. It cannot, however, be assumed that TLC-FID will be able to compete with already existing, completely automatic methods for the analysis of amino acids by ion-exchange chromatography[187].

2.1.6 Terpenes and Resins

Terpenes form a large group of organic substances present primarily in plants. Their molecules consist of two or more isoprene units (C_5) and are divided in dependence on the number of units into terpenes (C_{10}), sesquiterpenes (C_{15}), diterpenes (C_{20}), triterpenes (C_{30}), tetraterpenes (C_{40}) and polyterpenes. These materials are isolated from plant tissues and fluids (oils) either by extraction with suitable organic solvents or by steam distillation.

The selection of solvent systems for the chromatography of terpenes is based primarily on the type and number of polar constituents. Terpenes also include hydrocarbons, alcohols, ketones, acids and their esters. Groups with similar polarity can be separated on the basis of their size and the structure of the whole molecule.

So far, only a few papers have been published on the separation of terpenes and related substances on Chromarods. One of the reasons for this is probably the volatility of the low molecular weight terpenes, leading to low, irreproducible FID response.

Of the monocyclic terpenes, a mixture of menthol and its derivatives have been separated using a benzene-ethyl acetate-formic acid system (43:5:2) on Chromarods S II with quite good separation of menthol (R_F 0.66) from menthylglycoside (0.50) and menthylacetate (0.84)[188].

Of the resinous substances, resin has been separated from arnica (benzene-chloroform, 67:33) and tolu balsam (benzene-chloroform-formic acid 69:29:2), but the results obtained were not satisfactory. Resinoids from lichens (Eichenmoss) are separated better by using pure benzene on Chromarods S II[189].

Chromatography on Chromarods S II in an n-hexane-diethyl ether system yields good separation of the main components of a petroleum ether extract of the plants Petasites hybridus and P. albus. Novotný et al.[190, 191] have devoted a considerable amount of time to the analysis and structural clarification of this group of natural substances. It consists of a complex mixture containing sesquiterpenic compounds of the furoeremophilane (3, 4a β, 5β-trimethyl-4, 4a, 5, 6, 7, 8, 8a, 9-octahydronaphtho (2, 3-H)furan) group in addition to hydrocarbons, phytosterols, triterpenic alcohols and polymeric polar substances.

A characteristic component of the extract of P. hybridus is furanopetasin - (2-angenyloxy-9-hydroxyfuroeremophilane); the extract of P. albus contains primarily petasalbine (6-hydroxyfuroeremophilane) and albopetasine (6-angenyloxyfuroeremophilane). The identification of further components is evident from Figure 33.

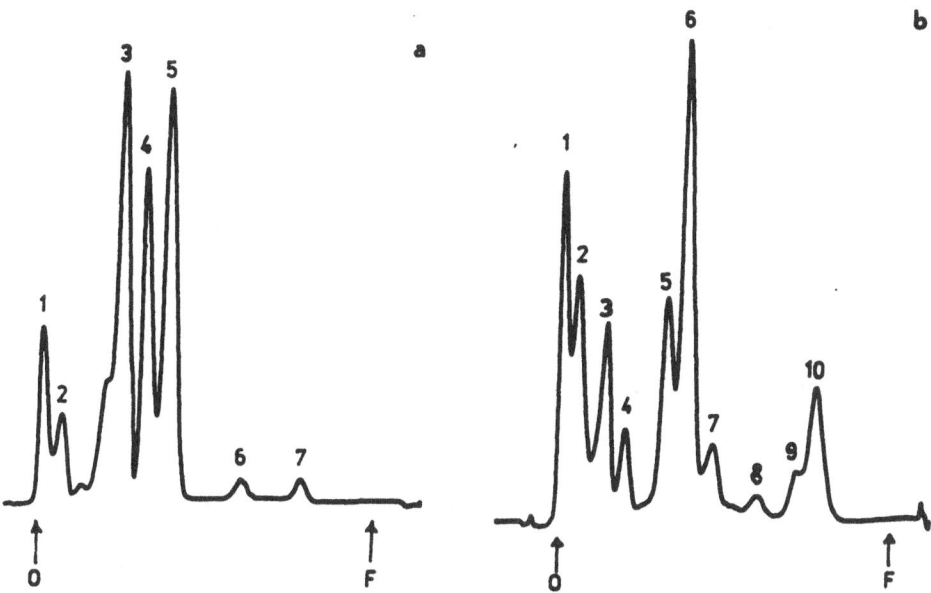

Fig. 33. TLC-FID of the petroleum ether extract of roots of Petasites hybridus (a) and P. albus (b)[184]. Chromarods S II, n-hexane-diethyl ether (82.5:17.5), scanning speed 0.31 cm s^{-1}. Peak identification: (a) (1), (2) polar components; (3) furanopetasine; (4) phytosterols; (5) 9-hydroxyfuroeremophilane; (6) esters of phytosterols; (7) hydrocarbons. (b) (1)–(4) unidentified polar compounds; (5) phytosterols and triterpenic alcohols; (6) petasalbine; (7) albopetasine; (8) phytosterol esters; (9), (10) hydrocarbons and furoeremophilane

Peak 2 in Figure 33(a) corresponds to 2, 9-dihydroxyfuroeremophilane. The polar substances (peaks 1–4) contained in P. albus include angenyljaponicin (3-angenyloxy-6-hydroxyfuroeremophilane) and petasolids.

The separation of a number of terpenic compounds was also described in the section on vitamins.

2.1.7 Psychotropic Substances

Psychotropic substances (psychopharmaceutics) are substances affecting the mental condition of the patient. They include substances decreasing mental activity (ataractica, neuroleptics, epileptics such as chlorpromazine, reserpine, diphenylmethanol derivatives, thioridazin, derivatives of propanediol such as meprobamate, etc).

The determination or separation of these substances has recently received a great deal of attention, especially in the control of drug levels among athletes and other sportsmen and in the analysis of pharmaceutical products.

Yamagushi et al.[192] applied the TLC-FID method to the determination of some psychotropic substances in the urine of sportsmen. A large number of

Table 32 TLC-FID of Psychopharmaceuticals

(a) Solvent system

No.	System	(v/v)	Reference
1	chloroform-methanol	93:7	192
2	benzene-acetone	75:25	192
3	chloroform-acetone	89:11	193
4	cyklohexane-diethyl ether-acetic acid-methanol	65:25:9:1	37
5	methanol-conc. ammonium hydroxide	99.1:0.9	188
6	methanol-conc. ammonium hydroxide	99.5:0.5	188

(b) Detection limits and R_F values

Drug	R_F in system		Detection limit μg
	1	2	
diazepam	0.75	0.67	0.05
chlordiazepoxid	0.65	0.30	0.25
oxazepam	0.60	0.42	0.25
serenace	0.36	0.21	0.25
defekton	0.36	0.18	0.20
triperidol	0.45	0.21	0.20
oxazolan	0.88	0.72	0.25
chlorpromazin	0.45	0.16	0.25
surmontol	0.51	0.30	0.30
oxypertin	0.72	0.30	0.50
	3	4	
meprobamat	0.17	0.23	
triphenylbenzene[a]	0.64	–	
methocarbamol[a]	–	0.15	

	5		6
thioridazin	0.28	chlorpromazin	0.25
sulforidazin	0.47	demethylchlorpromazin	0.36
thioridazinsulfoxide	0.63	dedemethylchlorpromazin	0.56

a internal standard.

elution systems were tested and mixtures of chloroform-methanol and benzene-acetone were found to be most effective (Table 32). An amount of 10 ml of urine was taken for analysis and sulphates and chlorides were decomposed using 0.5 ml of 5N sodium hydroxide. The mixture was then extracted using three portions of 5 ml of diethyl ether. The combined extracts were dried with sodium sulphate and filtered. The filtrate was evaporated at laboratory temperature in a stream of nitrogen. The residue was dissolved in 0.5 ml chloroform and 2 µl were applied to each Chromarod S. Both elution systems were employed for better separation of critical pairs (nos. 1 and 2 in (Table 32)). Very different detection limits were found. For example, diazepam can be determined in an amount of 0.05 µg, while oxypertin yields a perceptible response only at a rather large concentration (0.5 µg).

Other Japanese authors successfully employed the TLC-FID method for the determination of meprobamate in tablets, which (120 mg) were reduced to a powder and shaken with acetone (5 ml). The suspension was mixed with an internal standard (1,3,5-triphenylbenzene, 5 ml of a 0.6 % solution in acetone) and filtered; the solution was then applied to Chromarods S (1 µl) and chromatographed using system no. 3 (Table 32). The meprobamate content was found from a calibration curve. The reproducibility of the determination was reported to be very good (coefficient of variation 2.1 %). The yield was also satisfying (100.9 ± 2.1 %)[193].

Van Aerde, van Severen and Breackman[37] examined a similar problem. They selected methocarbamol (2-hydroxy, 3-(2-methoxy)phenoxypropyl ester of carbamic acid) as an internal standard and an elution system consisting of a mixture of glacial acetic acid, diethyl ether and methanol (Table 32). This procedure has the disadvantage that the mobilities of the internal standard and meprobamate are very similar; nonetheless, the reproducibility of the determination is very good (coefficient of variation 3.7 %) and could be improved by automatic sample application on rotating chromatographic rods.

2.1.8 Antipyretics, Analgesics and Hypnotics

In addition to the alkaloids of quinine, the most important substances for decreasing fever during illness (antipyretics) are the derivatives of salicylic acid (Aspirin), pyrazolone (Antipyrin) and p-aminophenol (Phenacetin). Some of these are analgetics, i.e. a group of pharmaceuticals decreasing the activity of the central nervous system (morphine, codeine). Typical hypnotic drugs that calm the nervous system are derivatives of urea, e.g. barbital and diphenylhydantoin, and include the sedative primidon.

So far, no systematic study has been carried out on the analysis of a mixture of these pharmaceuticals by the TLC-FID method. Table 33 lists the available information[181, 188].

Table 33 TLC-FID of Mixtures of Antipyretics, Analgesics and Hypnotics and Similar Substances
(a) Solvent system

No.	System (v/v)	Chromarods	Reference
1	chloroform-methanol-water 90:9.7:0.3[a]	S	181
2	acetone 100	S	181, 188

a chamber saturated with ammonia vapour.

(b) R_F values

System 1	R_F	System 2	R_F
diphenylhydantion	0.52	aminopyrine	0.36
phenobarbital	0.37	barbital	0.88
primidon	0.69	bromvalerylurea	0.82
		phenacetin	0.62
		isopropylaminopyrine	0.71
		caffeine	0.15

2.1.9 Antibiotics

Attempts to analyse antibiotics have been carried out by using Chromarods S II pretreated in two different acids[194]. The rods used for assaying tetracyclines were activated by immersion in chromic acid mixture over night and then washed thoroughly. Those Chromarods used to separate penicillin, chloramphenicol macrolides and aminoglycosides were pretreated with nitric acid. The authors limited their experiments to four mixtures of standard antibiotics, which they separated by using four different solvent systems (Table 34).

Table 34 TLC–FID of Antibiotics[194]
(a) Solvent system

No.	System
1	chloroform – methanol – 17 % ammonia 68 30 2
2*	chloroform – methanol – 17 % ammonia 34 33 33
3*	chloroform – pyridine-isopropyl acetate – water 27 9 18 46
4	chloroform** – methyl ethyl ketone** – pyridine – ethyl acetate 22 11 11 56

* upper layer; ** saturated with water.

(b) R_F values

System 1	R_F^a	System 2	R_F^a
benzyl-penicillin	0.40	dihydrostreptomycin	0.00
chloramphenicol	0.55	fradiomycin	0.40
		kanamycin	0.50

System 3	R_F^b	System 4	R_F^a
chlortetracycline	0.35	spiramycin	0.38
oxytetracycline	0.45	tylosin	0.51
		kitasamycin	0.58

Chromarods S II, pretreated:
a with nitric acid.
b with chromic acid mixture and washed with water before blank scanning.

2.1.10 Sulphonamides and Sulphanilic Acid

Only those compounds used in medicine as a group of antibacterial substances will be mentioned here. These substances are derivatives of sulphanilides differing in the type and the number of polar constituents and can be separated quite well using chromatography with a solvent system with high elution strength.

Okumura et al.[176] published a review dealing with the applications of TLC-FID to the analysis of pharmaceuticals. They give an example of the separation of sulphonamides using a polar system containing butanol, ethanol and acetic acid. Other Japanese authors separated the same mixture using a system containing a large amount of ammonium hydroxide[195]. As mentioned above, a high concentration of ammonia in the system usually leads to high background noise

Table 35 TLC–FID of Sulphonamides and Sulphanilic Acid on Chromarods S

(a) Solvent system

No.	System v/v	Reference
1	n-buthanol-ethanol-0.1 N acetic acid 60:20:20	176
2	ethyl acetate-ethanol-25 % ammonium hydroxide 63:21:16	195

(b) R_F values

Substance	System	
	1	2
4-(2-aminoethyl)-benzenesulphonylamide	0.09	0.82
sulphanilic acid	0.35	0.34
sulphacetamide	0.63	0.44
3-sulphenylamidoisoxazole	0.89	0.61

and also considerably decreases the lifetime of silica gel layers. Table 35 gives examples of the separation of some sulphonamides.

2.1.11 Polyamines

Putrescine (1, 4-diaminobutane) is formed in the decomposition of dead plant and animal tissues as a result of the action of anaerobic bacteria. TLC-FID analysis has demonstrated that this substance is present along with spermidine and spermine in extracts from malignant tumors. The extract was applied to Chromarods S II and eluted in a suspended developing tank (Figure 8) using a diethyl ether-chloroform (3:2) system in the metal reservoir and an aqueous ammonia system in the glass chamber. The rods were suspended for 10 min in the ammonia atmosphere of the chamber prior to immersion in the elution system. This procedure led to good separation and a very low noise level in the FID. It was thus possible to determine very small amounts of the individual polyamides in the sample (hundredths of microgram). The mobility on the layer was inversely proportional to the number of amino groups in the molecule and decreased in the order putrescine (R_F 0.41) > spermidine (0.33) > spermine (0.24)[196].

2.2 Applications in the Food Industry and Related Fields

As in biology and pharmacy, TLC-FID is also used in the food industry for the analysis of fats and their derivatives, i.e. acylglycerols and fatty acids. In contrast, applications in other areas in this industry are in their infancy and are limited to a few food additives and saccharides.

This section also includes work dealing with the analysis of cosmetics and perfumes, as cosmetics are part of the same branch of the fat industry.

2.2.1 Fats, Oils and Related Substances

Acylglycerols such as triacylglycerol, both positional isomers of diacylglycerols and monoglycerols and the fatty acids are technically the most important lipids. These are substances that are relatively easily to separate on thin layers, and consequently TLC-FID is one of the most important analytical methods in this branch of industry. It is used both in research laboratories and in the quality control of products in series analysis.

TLC-FID, so far, has been used in a large degree in the study of the glycerolysis of triacylglycerols and in the analysis of the products of the molecular distillation of monoacylglycerols. These substances are used extensively in the production of emulsified edible fats and especially as additives to bakery products, where they are used together with other substances, e. g. phospholipids. They are used to increase the bulk of bread and to act as preservatives.

The analysis of acylglycerols by classical chemical and physical methods, such as the determination of 1-monoacylglycerols by periodate oxidation[197], extractive separation[198] and separation of the products formed with urea[199] are now of limited importance. The routine control of the production processes and product quality are carried out using chromatographic methods alone, either column chromatography on a silica gel column (LC)[200, 201], gas chromatography[202, 203], high performance liquid chromatography[204, 205] or, recently, TLC-FID.

Column chromatography does not require expensive instrumentation but is tedious and time-consuming. A single worker can analyse only two to four samples during an eight hour shift. Glycerol and the free fatty acids are not separated well on a silica gel column and must be separated by other methods[201].

It is difficult to separate a mixture of acylglycerols using GLC because of their low volatility. It is, however, relatively simple to separate their trimethylsilyl derivatives. The determination of triacylglycerol is, however, rather difficult because it is eluted at high temperatures that have a detrimental effect on the quality of the stationary phase and also on the efficiency of the column[203].

So far HPLC has only limited applications in the routine analysis of acylglycerol emulsifiers. Not only is the instrumentation expensive, but also a

Table 36 Average Results of 10 Analyses of Product Samples of Molecular Distillation of Monoacylglycerol by the LC, GLC and TLC-FID methods

	Triacylglycerols			Fatty acids		
Method	LC	GLC[a]	TLC–FID	LC[b]	GLC	TLC–FID
$\bar{X}^c$	0.26	0.26	0.21	1.07	1.11	1.11
S D	0.19	0.19	0.10	0.07	0.12	0.06
C V %	73.1	73.1	47.4	0.6	10.9	5.5

	Diacylglycerols			Monoacylglycerols		
Method	LC	GLC	TCL–FID	LC	GLC	TLC–FID
$\bar{X}^c$	4.46	2.85	2.51	94.21	95.78	96.17
S D	0.25	0.20	0.11	0.38	0.38	0.28
C V %	5.6	7.1	4.5	0.4	0.4	0.3

a values taken from LC; b determined by titration; c average mass %.

suitable means of detection of the products of gradient elution remains pro-blematical[204].

The chief advantage of TLC-FID over the other separation methods is the short analysis time with approximately the same reproducibility, and the possibility of determining all the components in a single measurement. Comparing of the analysis of commercial molecularly distilled monoacylglycerols by LC, GLC and TLC-FID methods indicates that the results agree quite well, especially the GLC and TLC-FID analyses (Table 36).

A mixture of acylglycerols and fatty acids can best be analysed by the TLC-FID method using Chromarods S or S II. Chromarods A have been used only occasionally.

The main non-polar components of the elution system are either n-hexane (or petroleum ether) and benzene. The polar components are usually diethyl ether (combined with hexane) or ethyl acetate and chloroform (with benzene). Table 37 gives details of the compositions of systems tested.

Systems containing benzene can be used for a longer time because of their

Table 37   TLC-FID of Acylglycerols, Fatty Acids and Related Substances
(a) Solvent system (vol. %)

System no.	Benzene	Formic acid	Acetic acid	n-hexane	Diethyl ether	Acetone	Ethyl acetate	Chloroform	Toluene	Water	Methanol
1			3	94		3					
2		1.6		65.6	32.8						
3		1.0		59.4	39.6						
4	97		0.8				2			0.2	
5	96.4		0.7				2.7			0.2	
6	96	1.3					2.7				
7	95.6		1.0				3.0			0.4	
8	95.0		2.0				3.0				
9(a)	90.8	0.3				4.3	4.3				0.3
9(b)	99.0	0.1				0.9					
10	75.4	1.0		22.8							0.8
11	69.2	2.0						28.8			
12	68.6	2.0						29.4			
13	67.0	1.8						31.2			
14	58.8	2.0						39.2			
15	39.2	2.0						58.8			
16			0.8				2.0		97.0	0.2	
17		1.8		12.5	3.5			26.2	56.0		
18(a)				80	20						
(b)								44.4	44.4		11.2
(c)		3.5			8.8			52.6	35.1		
19*								100			

Note: system 9(a) to 5 cm; 9(b) to 10 cm, system 18(a) to 12 cm, 18(b) to 6 cm, 18(c) to 3 cm, * presaturation with chloroform.

(b) R_F values

System no.	Chroma-rods	MG	DG		FA	TG	Reference
			1.2	1.3			
1	S	0.17	0.26	0.32	0.57	0.66	36
	S II	0.15	0.26	0.31	0.65	0.76	
2	S	0.20	0.53	0.60	0.66	0.75	36
	S II	0.20	0.53	0.68	0.73	0.87	
3	S	0.10	0.39	0.45	–	0.62	207
4	S II	0.10	0.25	0.34	0.51	0.67	206
5	S II	0.10	0.34	0.43	0.59	0.75	36
6	S	0.16	0.35	0.41	0.55	0.68	36
	S II	0.18	0.38	0.46	0.61	0.74	
7	S	0.20	0.42	0.47	0.56	0.63	36
8	S	0.19	0.47	0.56	0.67	0.79	36
9	S	0.20	0.41	0.47	0.54	0.61	36
	S II	0.22	0.51	0.58	0.65	0.78	
10	S II	0.08	0.34	0.46	0.60	0.78	36
11	S II	0.18	0.40	0.58	0.69	0.78	36
12	S	0.10	0.36	0.45	0.57	0.70	116
13	S II	0.06	0.30	0.43	0.58	0.78	208
14	S	0.08	0.34	0.46	–	0.72	209
15	S	0.14	0.57	0.67	–	–	210
16	S II	0.15	0.37	0.42	0.53	0.63	36
17	S	0.10	0.38	0.48	0.58	0.73	36
18*	A	0.29	0.48	0.48	0.13	0.78	211
19	S	0.09	0.25	0.25	–	0.44	36

* system 18 also separates the methyl esters of the fatty acids (R_F 0.88).

lower volatility (usually up to five times) without apparent changes in their separation abilities. They are, however, more toxic; the toxicity can be decreased somewhat by replacing benzene with toluene (system no. 17, Table 37). Elution mixtures containing larger amounts of n-hexane are relatively sensitive to temperature changes in the laboratory. They are, however more easily removed from the elution layer than benzene or toluene systems; the effect on the background noise in the FID is negligible. Low background noise in detections can also be attained by using systems containing benzene and toluene that contain acetone and ethylacetate (nos. 4–9) when the eluted Chromarods are dried for 7 min at 80 °C[36].

Relatively good separation can be attained using systems based on a combination of benzene with chloroform (nos. 11–15). However, monoacylglycerol is too close to the start, the elution times are about 20 % longer than for the other systems and the eluted rod must be dried for a longer period of time. The presence of organic acids (formic, acetic) ensures that the fatty acids yield regular, narrow zones without tailing. Formic acid is usually preferred; it is less soluble in hexane and benzene systems than acetic acid but is more easily removed from the thin layer during drying prior to scanning in the detector. The R_F value for fatty acids is lowest on Chromarods A and the positional isomers of diacylglycerol are not separated (system no. 18). An increase in the polarity of the system did not produce any improvement. A small amounts of methanol added to benzene and n-hexane systems (no. 10) increases the mobility of monoacylglycerol, which would otherwise remain at the start. Larger amounts of methanol in the system lead to merging of the fatty acid peaks and triacylglycerol and increase the detector noise. Replacing benzene by toluene (compare systems nos. 4 and 16) increases the mobilities of mono- and diacylglycerols but leads to lower separation of the two diacylglycerol isomers.

Simple acylglycerol mixtures can be separated by elution using chloroform alone, without ethanol (system no. 19), combined with 15 min preconditioning in the same solvent. This chromatographic procedure yields symmetric peaks

Table 38 Dependence of the Correction Factors ($K_F \times 10^2$, reciprocal response) and Reproducibility on the Type of Solvent System and Chromatographic Layer (S and S II). Average of 10 determinations. Palmitic acid and its derivatives. Scanning speed 0.41 cm s^{-1}, H$_2$ flow-rate 180 ml min^{-1}, air 2.1 l min^{-1} (References 36,212)

(a) Correction factors

System no.	MG		1.2 DG		1.3 DG		FA		TG	
	S	S II	S	S II	S	S II	S	S II	S	S II
1	96	88	92	94	100	97	105	107	109	121
2	97	90	91	94	105	102	96	95	114	127
4	97	86	94	92	102	97	96	101	111	142
5	91	73	92	91	103	106	96	132	129	120
6	94	81	92	95	100	108	93	106	128	122
7	94	84	94	98	102	112	96	91	115	123
8	85	74	89	92	101	106	103	110	135	153
9	97	92	97	92	93	92	107	99	108	110
10	102	93	90	94	85	97	107	105	122	110
11	106	86	114	97	102	112	95	107	115	100
16	103	96	92	91	103	111	103	92	102	118
av.*	97	86	94	94	100	104	100	104	117	122

* average

(b) Reproducibility of the determination of the correction factors (C V %)

System no.	MG		1.2 DG		1.3 DG		FA		TG	
	S	S II	S	S II	S	S II	S	S II	S	S II
1	3.5	7.9	2.6	2.7	2.3	12.5	4.7	8.5	4.0	11.5
2	4.8	5.9	2.0	5.2	2.0	3.5	3.1	3.8	2.6	3.9
4	3.7	4.9	1.5	4.9	2.7	3.2	3.4	10.5	5.4	9.3
5	5.5	16.7	6.1	9.5	4.2	8.2	2.0	12.8	5.2	15.3
6	4.1	6.2	1.7	3.2	1.3	4.2	2.0	6.4	3.8	10.7
7	3.7	2.6	2.5	4.5	3.0	5.5	1.7	2.2	2.9	6.9
8	0.9	5.1	2.1	2.3	0.9	6.3	3.9	2.0	3.8	12.5
9	7.6	4.7	4.6	2.3	1.3	3.4	6.9	1.3	3.5	2.6
10	2.4	3.2	4.9	4.0	6.2	3.3	4.0	3.7	7.3	6.2
11	4.4	8.4	9.9	3.0	2.1	4.1	2.2	5.4	4.8	6.2
16	1.9	2.6	2.1	2.2	3.1	2.0	2.5	3.3	3.4	4.9
av.	3.8	6.2	3.6	4.0	2.6	5.1	3.3	5.4	4.2	8.2

(c) Reproducibility of R_F values (C V %)

System no.	MG		1.2 DG		1.3 DG		FA		TG	
	S	S II	S	S II	S	S II	S	S II	S	S II
1	0.9	0.6	1.5	1.8	2.0	2.1	1.9	4.7	2.8	5.4
2	1.3	1.5	2.0	2.8	2.5	2.3	3.8	3.9	5.4	8.2
4	1.6	1.3	1.0	2.2	0.9	2.3	1.3	2.6	3.4	8.6
5	1.0	1.2	1.6	1.7	2.1	2.2	2.4	3.0	4.6	6.2
6	0.7	0.9	0.8	1.1	1.6	1.7	2.4	2.4	4.3	4.9
7	0.9	1.1	1.6	2.0	2.3	1.9	2.1	1.7	4.6	4.1
8	0.8	1.6	1.5	2.1	2.5	2.4	2.1	2.5	4.9	3.4
9	1.0	1.0	1.5	2.3	2.8	3.5	2.3	6.5	5.3	14.9
10	1.0	0.8	1.6	1.9	2.2	3.6	1.5	3.3	5.5	10.8
11	0.6	1.0	1.6	2.7	1.8	1.6	1.3	4.5	2.7	4.5
16	0.7	2.0	0.9	1.1	0.8	1.3	2.0	1.8	2.8	2.8
av.	0.9	1.2	1.4	2.0	2.0	2.2	2.3	3.4	4.1	6.7

Solvent systems in Table 37.

and very good reproducibility (coefficient of variation about 2 %)[36]. As for phospholipids (see Section 2.1.1), the content of water in the activated layer plays an important role in the TLC-FID of acylglycerols and fatty acids. Too high a water content leads to partial overlap of the peaks of 1, 3-diacylglycerol and the fatty acids; too highly activated layers yield poor separation of the two diacylglycerols and of fatty acid and triacylglycerol (system no. 12)[116].

The reproducibility and responses for the individual lipids are discussed in a number of papers[31, 55, 126, 205 206]. The coefficient of variation ranges from 1 to 20 %, depending on the amounts of the individual lipids applied.

Recently published results have shown that aliphatic unsaturated lipids exposed to iodine give greater responses than unexposed lipids[55] (see also Paragraph 1.2.3.6).

Both the responses and reproducibilities depend to a certain degree on the type of thin layer and elution system used (Table 38). The experimentally determined average responses decrease with the R_F value of the given lipid (i. e. the correction factors increase). For example, triacylglycerol produces the lowest response even though it contains the largest number of carbons. The reproducibility determined for both types of layers is clearly best achieved by using elution system no. 16 (with toluene, and with the similar composition as no. 4, with benzene). The separation of the components at both the start and the front is very good and the background noise level is very low. It follows from the average reproducibility values that the error in determinations on Chromarods S is much lower than that obtained using the more expensive S II rods (i. e. only 2 –3 %).

This is also true of the reproducibility of the R_F value (Table 38). The greatest scatter was found for the most mobile substance, triacylglycerol.

The correction factors are constant for all the acylglycerols and fatty acids, in the range 0.5–10 µg. i. e. the calibration curves for the dependence of the response on the concentration are linear and pass practically through the origin[212].

On the other hand, Chromarods S II are irreplaceable in some separations. For example, they yield better separation of the acylglycerols of isomeric fatty acids. Ackman[35] gives an example of such a separation. Chromatograms of mixtures of diacylglycerols and triacylglycerols of octadecane and 16-methylheptadecane acids, obtained by elution with an n-hexane-diethyl ether-acetic acid system (80 : 6 : 0.5), are compared on the two types of silica gel layers. The difference in the separation ability is considerable. While Chromarods S yield only six incompletely resolved peaks, S II rods yield eight much narrower peaks.

The positional isomers of monoacylglycerol are not separated either on silica gel or on alumina layers. Tanaka et al.[31,81], however, state that good separation can be obtained (as in classical TLC) using silica gel impregnated with boric acid. The Chromarods S II are immersed for 5 min in an aqueous 3 % solution of boric acid and dried for 5 min at 120 °C, followed by activation in the FID. A chloroform-acetone mixture (96 : 4) is used as the elution system and separates all the components on a 10 cm path length, i. e. 1-monoacylglycerol (R_F 0.02), 2-monoacylglycerol (0.16), fatty acid (0.35), 1, 2-diacylglycerol (0.45), 1, 3-diacylglycerol (0.53) and triacylglycerol (0.65). It is interesting that fatty acid migrates between 2-monoacylglycerol and 1, 2-diacylglycerol. Its mobility is, however, increased to such an extent when a chloroform-acetone-acetic acid system (100 : 1 : 1) is used that it lies between the two diacylglycerols. This method has been recently used for the separation of the both positional monoacylglycerol isomers in a lamellar crystalline phase with excess water[213].

Tatara et al. used Chromarods S II impregnated with boric acid to separate products of the hydrolysis of the triacylglycerols by pancreatic lipase[59]. When an n-hexane-diethylether-acetic acid system (70 : 30 : 0.1) is used, an internal standard (p-carbethoxybenzylalcohol) can be added to the sample; this substance migrates between 1-monoacylglycerol and 1, 2-diacylglycerol. In this system, fatty acid has the "normal" position between 1,3-diacylglycerol and triacylglycerol. All the components are evenly distributed over the whole length of the layer; the peaks are symmetrical and have almost identical widths, improving the reproducibility of the determination. The relationship between the peak ratio A_i/A_{st} and the ratio of the corresponding masses M_i/M_{st} (see Section 1.2.3) had the form

$$M_i/M_{st} = 0.843 \, A_i/A_{st}^{0.678} \ (r = 0.992)$$

The layers can be used at least five times without renewing the impregnation. After loss of separating ability they are immersed for 12 h in concentrated nitric acid and washed several times with distilled water. They are then dried for 1 h at 120 °C and impregnated once again in the same manner.

Silver nitrate chromatography[214] is very useful for verifying lipid structures. Development of this procedure led to TLC-FID using Chromarods S II impregnated with a silver nitrate solution[30, 81, 215—222]. The Iatron company recommends this procedure[221, 222]: Impurities are removed by combustion in the FID, the rod is immersed for 5 min in a 5 % aqueous solution of silver nitrate in a test tube protected from the light by aluminium foil. The impregnated rods are then dried for 30 min at 120 °C in a dark drying box, and 20–30 μg of the sample is applied. The rod is then eluted in a tank (wrapped in aluminium foil) either in a benzene-diethyl ether (97 : 3) or a benzene-chloroform-acetic acid (90 : 10 : 1) system. A neutral system is suitable for separating a mixture of saturated and unsaturated lipids with a smaller number of double bonds; a system containing acetic acid is used for separating unsaturated vegetable oils. The eluted rods are dried for 5 min at 120 °C and detected immediately in the FID. Prior to the next analysis, the rod must be immersed (preferably overnight) in concentrated nitric acid and the acid removed by washing with distilled water. The rod can be reimpregnated after drying for 1 h at 120 °C.

The lifetime of rods used in argentation chromatography is somewhat reduced (about ten analyses)[218, 221].

The mobility of the triacylglycerols on impregnated layers increases with the saturation of the bonded fatty acid. For example, the R_F values or tripalmitin, triolein, trilinolein and trilinolenoin in a benzene-diethyl ether system (97 : 5) correspond to 0,73, 0.62, 0.44 and 0.21, respectively[223].

Argentation TLC-FID has been proposed as a fast method for the identification of fats and oils, e. g. cocoa butter, palm and coconut oils, beef tallow, olive oil and soya, rape, flax, corn and safflower oils[81, 215, 221]. Figure 34 gives an example of such an analysis.

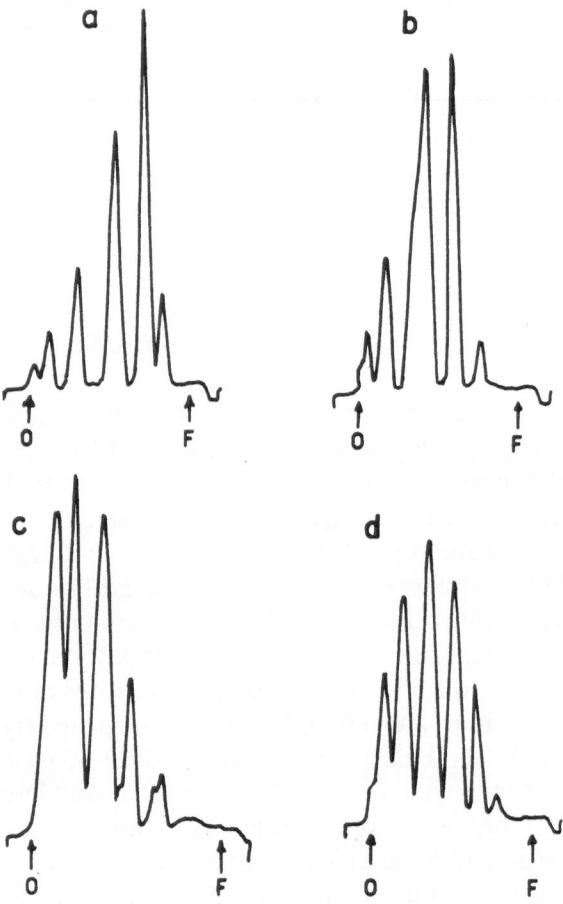

Fig. 34. The TLC-FID of plant oils on Chromarods S II impregnated with silver nitrate using a benzene-diethyl ether system (97:3, A) or benzene-chloroform-acetic acid (100:10:1, B). (a) palm oil. system A; (b) olive oil, system B; (c) sunflower oil, system B; (d) corn oil, system B, H_2 flow-rate 160 ml min^{-1}, air 2 l min^{-1}, scanning speed 0.41 cm s^{-1} [221]

Ackman and Sebedio[218, 219, 224] used argentation TLC-FID for the study of the geometric isomerization of the methyl esters of unsaturated fatty acids obtained from partially hydrogenated fish oils. When elution using benzene alone is employed, the isomers are separated into three groups; least polar are the cis-cis dienes with adjacent double bonds, followed by a mixture of cis-cis dienes with one methylene group between the double bonds and cis-trans and trans-cis dienes with double bonds in neighbouring positions, and finally by a mixture of cis-trans and trans-cis dienes and trans-trans dienes with double bonds separated by a single methylene group. They used a 2.5 % silver nitrate solution in acetonitrile for impregnation of the Chromarods S and activated the rods for 3 h at 120 °C. Elution was carried out using benzene or an n-hexane-benzene system (1 : 1).It is interesting that the correction factors for the cis-cis isomers were larger (1.33) than for the corresponding trans-trans isomers (0.98); i.e. the opposite to that found for GLC-FID[219].

The cis and trans isomers of fatty acids with one or more double bonds can be separated in a similar manner. The system described above can be used to separate the methyl esters of unsaturated acids in the following manner: 18 : 3-9, 12, 15 cis, cis, cis (R_F 0.15), 18 : 2-9,12 cis, cis (R_F 0.29), 18 : 1-9 cis (0.47), 18 : 1-9 trans (R_F 0.62) and 18 : 0 (R_F 0.78). All the components were separated to the base line, which has a very low noise level. The mobility was again inversely proportional to the number of double bonds, and the trans isomers were adsorbed more strongly then the cis derivatives[219].

Inconveniences connected with the preparation of impregnated layers and with their decreased lifetimes were avoided by Petersson[225] by converting the unsaturated triacylglycerols to the mercuric acetate derivatives, which could then be separated on ordinary Chromarods S using a chloroform-petroleum ether-acetic acid-methanol system (25 : 25 : 1.5 : 0.15-0.40). The methanol content in the system was adjusted according to the rod age and the concentration of the oil analysed. Systems with higher methanol concentrations were separated better than the unsaturated components. However, the separation of the saturated triacylglycerols from the adducts of the unsaturated derivatives was inferior.

The procedure for the preparation of the adducts is quite simple[225]. About 10 mg of sample is dissolved in 300 µl of benzene and 300 µl of methanol and mixed with 40 mg of mercuric acetate. The mixture is then refluxed for 20 min. The solvent is evaporated and the residue is dissolved in chloroform and applied to the rods. The mobility of the adducts increases with the saturation of the initial triacylglycerols (R_F values: trilinolein 0.16, triolein 0.36, dioleinpalmitin 0.59, stearinoleinpalmitin 0.73, tripalmitin 0.88). The dependence of the response on the amount of sample applied is linear only in the range 0.25 to 3 µg. The adducts yield a response different from that of the initial triacylglycerols. For example, if the relative response of pure palmitin is 1.00, the corresponding response of the adducts stearinoleinpalmitin, dioleinpalmitin and triolein are 0.82, 1.05 and 1.34, respectively. This method has been used to analyse palm oil, shea fat and mango seed oil with quite good reproducibility (saturated triacylglycerols have a coefficient of variation of 11 %, adducts only 1.7–2.9 %).

In the fat industry TLC-FID has also been applied to the analysis of castor oil[226, 227], which is separated using an isopropyl ether-benzene-acetone system (10 : 10 : 1) on Chromarods S II to the triacylglycerol of ricinoleic acid (R_F 0.58) and a mixture of the remaining triacylglycerols (0.64). The literature contains a number of papers on the determination of lipids in yeast[81], in kangaroo meat[228] and poultry[229], in tomato fruit protoplasts[230] and especially in marine organisms[57, 109, 111, 144, 224, 231—245].

Fish fats contain acylglycerols (primarily TG) and free fatty acids in addition to phospholipids, cholesterol and its esters and hydrocarbons. Stepwise elution using petroleum ether-benzene-formic acid (92 : 17 : 1) and petroleum ether-diethyl ether-formic acid systems (97 : 4 : 1) on Chromarods S separated all the components almost evenly along the whole length of the rod in the order (from the start): phospholipids, cholesterol, triacylglycerol, fatty acids, sterol esters and hydrocarbons[244].

Problems encountered in the practical application of Iatroscan Chromarods in the analysis of lipids obtained from sea water are discussed by Parrish and Ackman[242, 243]. Stepwise development in systems of the HDF type with different elution strengths, combined with selective scanning of the less polar components permits separation of the lipids into basic classes according to their decreasing polarity (aliphatic hydrocarbons, aromatic hydrocarbons, waxes, methyl esters of the fatty acids, fatty ketones, acylated glyceryl ethers, fatty alcohols, sterols and polar lipids).

Kaitaranta and Ke[246] used the HDF elution system for studying the oxidation of fish oils obtained from mackerel, herring and tuna. The oxidation products are very polar and remain at the start together with the phospholipids. The increase of the signal at the start is proportional to the decrase in the area of the triacylglycerol signal.

The use of less polar systems, e. g. a mixture of toluene and n-hexane (1 : 1) yields very good separation of oxidized and unoxidized esters into two signals on Chromarods S. An example of such an application is the analysis of partially oxidized methyl esters of linoleic acid, was mentioned previously (Section 1.2.3). Suitable selection of two starting points permits a single rod to be used for the chromatographic analysis of two samples. (Figure 13). The direct proportionality of the area of the signal of the oxidized component and the peroxide number admits the possibility of applying TLC-FID in the evaluation of the oxidative deterioration of fats to replace the tedious chemical processes now employed[36].

HDF-systems have also been used for monitoring fatty acid polymerisation and isolation processes. Optimum elution strength of these types of solvent mixtures can be achieved by adjusting the ratio of diethyl ether and organic acid. For example, a mobile phase based of n-hexane (or n-pentane), containing 15 % by volume of diethyl ether and 1 % of formic (or acetic) acid exibits a very similar resolving power to system with only 3 % of ether and the same amount of formic acid. An elution mixture with lower diethyl ether content was found to be more sensitive to change in the composition as a result of the loss of ether produced by its high volatility and by its preferential sorption in the lower region of the thin layer. All the components form singlets with R_F 0.10 (trimers and higher oligomers), R_F 0.40 (dimers) and R_F 0.50 (monomers). The trimer fraction can be split into about four fused peaks by the second elution in system dichlormethane-methanol-acetic acid (98 : 0.5 : 1.5) over a short migration distance (about 2 centimetre above the starting point). This resolution profile could be used to confirm the identity of various commercial samples of higher oligomers[247].

As in biology (see Section 2.1.1), TLC-FID can be used in the fat industry to determine the overall fat content of raw materials and products by focussing the samples into narrow zones using a mixture of chloroform and methanol (1 : 1)[248].

2.2.2 Phospholipids

Interest in the application of phospholipids in foodstuffs as a result of their favourable biological and surface active properties is reflected in the development of methods for the analysis of these substances, which are widely distributed in nature. Most of the published papers and the applications of TLC-FID for quantitative analysis of natural phospholipids are connected with biological problems (see Section 2.1.1). This is probably because biologists have had the greatest experience with the use of classical TLC. So far, no papers have been published on the utilization of TLC-FID for the determination of individual phospholipids in food raw materials and products, except for the determination of phospholipids in egg yolk[175], in spinach[249] and in sea urchin yolks[250] and manufacturers' chromatograms of soya lecithin[175].

While no problems are encountered in the separation and quantitative determination of neutral (amphoteric) phospholipids using chloroform-methanol-water systems (see Section 2.1.1), the separation of acid groups of substances, such as a mixture of phosphatidic acid and polyglycerophosphates is rather difficult. These are substances that have recently become quite important as non-toxic emulsifiers for a number of applications in the food industry and other branches of industry[251,252]. These complex mixtures contain main components, i. e. the salts of phosphatidic acid (PA) and bis(diacylglycerol)phosphate (BDP), as well as other glycerophospholipids (salts of lysophosphatidic acid (LPA), bis(monoacylglycerol)phosphate (BMP) and monoacylglycerol-diacylglycerophosphate (MDP)) in addition to the products of side reactions (the salts of glycerophosphoric acids (GPA) and various polyphosphoric acids) and finally the starting substances (a mixture of acylglycerol and fatty acids). The separation of such a large number of substances, as well as a number of positional isomers, cannot be carried out on a single rod. However, from a practical point of view (to study the progress of synthesis and control of the final products), even limited information on the contents of the main components is useful. Analysis and densitometric quantification of mixtures of phosphatidic acids on classical silica gel layers can be carried out[253], but is not useful for routine control of the production process and quality control. These mixtures are incompletely separated on Chromarods S and S II. The individual components form rather broad, partly overlapping zones spread over the whole length of the chromatographic rod.

We have found that the best approach involves methylation of the sample using diazomethane. The methylated derivatives are separated quite well in less polar solvent systems and form sharp zones with low noise levels (Fig. 35)[122,212].

In addition, methylated phosphatidic and bis(diacylglycerol)phosphate cannot react with some divalent cations in the stationary layer producing com-

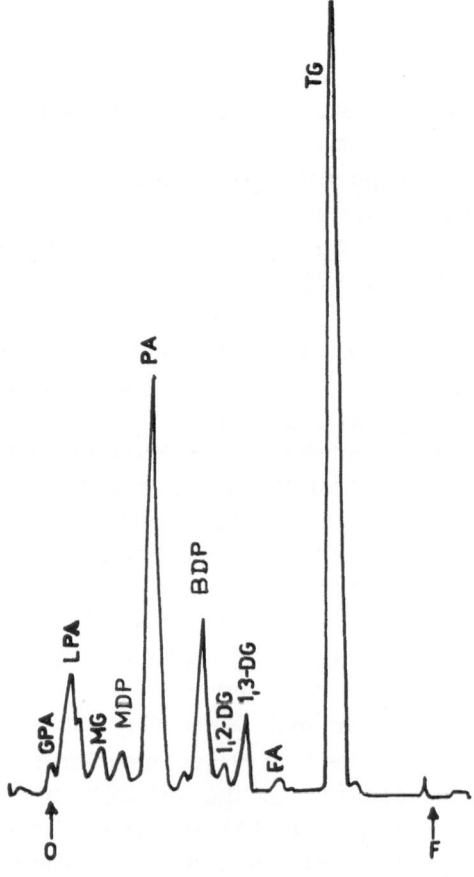

Fig. 35. TLC-FID of a mixture of phosphatidic acid and bis(diacylglycerol)phosphate (after methylation using diazomethane) on Chromarods S II using system no. 9, Table 39[122,212]

pounds that would be poorly combustible in the FID, which could interfere in subsequent analyses[122, 254].

Solvent systems for the separation of phosphatidic acids (nos. 1 and 2 in Table 39) contain chloroform and methanol in addition to ammonia, which is less favourable (see Section 2.1.1.). In contrast, solvents suitable for methylated phosphatidic acids (nos. 3 to 9) consist of highly volatile substances that have practically no effect on the behaviour of the thin layer in the FID. It is suggested that higher reproducibility can be attained during the separation by soaking the rods for several minutes in distilled water after each analysis and by reactivating them in the FID before application of the next sample[212].

The best separation of all the components, i. e. both methylated phosphatidic acids and neutral lipids, can be attained by stepwise elution in systems nos 9 (a) and 9 (b). The elution is somewhat more complex and time-consuming, but the individual peaks are sharp and well separated. In contrast to system no. 9, systems containing methanol are intended for the separation of the more polar

Table 39 TLC-FID of Mixtures of Phosphatidic Acids and Bis(acylglycerol)phosphates
(Phosphorylated acylglycerols)[122, 212]

(a) Solvent system
(a)(i) Unmethylated samples (acids or salts)

System no.	Chloroform	Acetone	n-Heptane	Methanol	28 % Ammonium hydroxide
1[a]	60	–	10	24	6
2	61	6	–	27	6

a   system for TLC on classical layers[253]

(a)(ii) Methylated samples

System no.	Benzene	Acetone	Ethyl acetate	n-hexane	Water	Formic acid	Methanol
3	75.8			22.7			1.5
4	75.3			22.6			2.1
5	75.1			22.5			2.4
6	74.6			22.4			3.0
7	71.4			21.4			7.2
8	62.5			18.7			18.8
9[a]	92.0	4.0	2.6		1.3	0.1	
9[b]	99.1	0.82				0.08	

9(a) developed twice for 5 min; 9(b) once for 30 min

(b) R_F values

System no.	LPA	PA	BMP	MDP	BDP	MG	DG		FA	TG
							1.2	1.3		
1	0.02	0.18	0.34	0.42	0.45	0.45	0.73	0.73	0.27	0.82
2	0.00	0.26	–	–	0.82	–	–	–	–	–
3	0.00	0.30	0.08	0.21	0.62	0.10	0.47	0.56	–	0.79
4	0.02	0.48	0.15	0.29	0.70	0.12	0.54	0.64	–	0.82
5	0.05	0.49	0.29	–	–	0.23	0.64	0.72	–	0.80
6(a)	0.21	0.62	0.33	0.62	0.80	0.25	0.69	0.75	–	0.80
7(b)	0.40	0.74	0.64	0.75	0.80	0.51	0.74	0.75	–	0.80
8(c)	0.40	0.79	0.63	–	0.80	0.62	0.80	0.80	–	0.80
9	0.05	0.25	0.06	0.19	0.34	0.11	0.42	0.50	0.58	0.78

(a) plus GPA – 0.02; (b) GPA – 0.12; (c) GPA – 0.22.

components (LPA, BMP, MDP and possibly also GPA). However, at higher methanol levels, neutral lipids are not separated from the methyl esters of phosphatidic and bis(diacylglycerol)phosphate, A system with 2.1 % by volume of methanol separates almost in the same way as system no. 9, except for the increased mobility of the methyl ester of bis(diacylglycerol)phosphate, which migrates between 1, 3-diacylglycerol and triacylglycerol. In addition, the separation of the components is very sensitive to even small changes in the methanol content in the system, so that system no. 9 is preferable for routine analyses.

The responses of the methyl esters of phosphatidic and bis(diacylglycerol)-phosphate in the FID are quite similar to the response for neutral lipids and do not depend on the analysed amount in the range 0.7–10 µg. However, the relative responses decrease with the scanning rate, i. e. with decreasing pyrolysis temperature. In contrast, the response of triacylglycerol increases under these conditions, indicating partial vapourization of triacylglycerol in the region

Table 40 Dependence of Correction Factors K_F (Reciprocal of the FID response) and Coeffitient of Variation of the Methyl Esters of Phosphatidic Acid, Bis(diacylglycerol)phosphate and Acylglycerols on the Scanning Speed. Average of ten determinations by system no. 9 on Chromarods S[212]

Scanning speed cm s^{-1}	K_F					C V^a %	C V^b %
	TG	1,3 DG	BDP	PA	MG		
0.21	1.64	0.83	0.94	0.90	0.99	6.7	–
0.31	1.27	0.84	0.97	1.05	0.96	2.9	–
0.42	1.07	0.89	0.99	1.09	0.97	3.0	2.5
0.58	1.00	0.89	1.17	1.20	–	3.2	1.2
0.73	0.89	0.89	1.28	1.15	0.93	3.6	2.3

a   CV % at flow-rates of H_2 180 ml min^{-1} air 2.1 l min^{-1}; and   b   CV % at H_2 250 ml min^{-1}, air 1,5 l min^{-1}.

outside the flame with slow velocity of the rod through the detector. The average coefficient of variation at medium detection rates is about 3 %, which can, however, be decreased by suitable adjustment of the detection conditions (Table 40)[212].

The optimal speed of movement of the rod through the detector is $0.42\,\mathrm{cm\,s^{-1}}$. At lower rates, rod overheating occurs and part of the more volatile components are lost (especially triacylglycerol). At higher velocities, (0.58 or $0.73\,\mathrm{cm\,s^{-1}}$), part of the separated substances remain uncombusted (about 5–10 % by weight)[122,212].

Phosphatidic acid and bis(diacylglycerol)phosphate form salts with most biogenic cations (e. g. salts of calcium, magnesium, iron, copper, cobalt, zinc, sodium and potassium). These salts can be used in food products and fodders such as roburancia[252]. They can be separated from the free phosphatidic acid and bis(diacylglycerol)phosphate on the basis of their different solubilities in acetone or ethyl acetate[36]. For example, the kinetics of the neutralization of phosphatidic acids can be studied on Chromarods S II after stepwise elution in systems containing benzene-acetone-ethanol-water ($30:40:10:2$, to 4 cm from the start), benzene-acetone-ethyl acetate-water-formic acid ($70:30:2:1:0.1$, to 5.5 cm) and HDF ($42:8:0.08$, to 10 cm). In the first system, all the components migrate with the front except the salts of phosphatidic acid and bis(diacylglycerol)phosphate which remain at the start. The second system ensures better separation of free phosphatidic acid and bis(diacylglycerol)phosphate (one peak) from diacylglycerol and the third system yields a separation of diacylglycerols (in a single peak) from fatty acids and triacylglycerol[36].

Good separation of the natural phospholipids in medicine and biology (Section 2.1.1) were found in a study of the effects of refining on the distribution of phospholipids in vegetable oils (corn, sunflower, peanut). The phospholipid fraction obtained by preliminary chromatography on a silica gel column was dissolved in chloroform and applied to Chromarods S II. The remaining simple lipids were removed from the rod using selective scanning after elution with acetone. The rods were then conditioned stepwise in chambers with 65 % relative humidity (10 min) and with the vapours of the elution system, chloroform-methanol-water ($69:28.5:2.5$) (10 min). Subsequent elution in this solvent system separated the phospholipids from the origin in the order lyso-phosphatidylcholine, phosphatidylcholine, phosphatidylinositol, phosphatidylethanolamine and phosphatidic acid. The FID responses decreased in the order: phosphatidylcholine > phosphatidylethanolamine > phosphatidylglycerol > phosphatidic acid > phosphatidylinositol. The phosphoholipid fraction from raw oils contained about 60 % phosphatidylcholine, 6–10 % phosphatidic acid, 10–20 % phosphatidylethanolamine and 14–20 % phosphatidylinositol. In the refining of raw oils, most of the phospholipids were contained in the hydration phase (gums)[255].

The overall phospholipids in the raw and hydrated (degummed) vegetable oils are usually determined gravimetrically after precipitation using acetone[64]. An alternative method is TLC-FID. About 1 g of oil is weighed in a centrifuge tube and is mixed with 10 ml acetone saturated with potassium iodide. The mixture is centrifuged at 800 g for 20 min. The supernatant is decanted and about 3 ml of acetone is added (with potassium iodide). The suspension is stirred and again centrifuged. After decantating, the sediment is dissolved in 1 ml of a chloroform--methanol mixture (2:1) containing 0.5 % by weight methylheptadecanoate (internal standard). An amount of 1 µl of solution is applied to Chromarods S or S II and developed for 45 min in a system of 1,2-dichloroethane-chloroform--acetone-acetic acid (94:5:1:0.1). The presence of acetone in the eluent prevents migration of the phospholipid fraction. A typical chromatogram contains the peaks of phospholipids (R_F 0.0)[256], fatty acids (0.50-tails) and the internal standard (0.75)[256].

2.2.3 Saccharides

This section deals with monosaccharides and oligosaccharides and their derivatives and some related substances.

The chromatographic behavior of most sugars depends on the presence of a large number of polar hydroxyls that form strong adsorption bonds with silica gel. Thus a solvent system with high elution strength is required for their migration and separation.

It was mentioned in the introduction that Chromarods can be developed in very polar solvents. It can thus be assumed that they can be used for the chromatography of very simple hydrophilic sugars. So far, however, only three chromatograms are available, for the separation of malto-oligosaccharides and isomalto-oligosaccharides on Chromarods S II in an ethyl acetate-formic acid-water system (6:3:1), unfortunately without identification of the individual zones[180]. Further, a mixture of rafinose (R_F 0.35), maltose (R_F 0.54) and rhamnose (0.70) was separated using an acetone-water (91.8:8.2) system on the same type of rods[188].

Attempts to separate more complex saccharides, steviosides, were quite successful. These are natural sweeteners obtained from the leaves of plants of the species Stevia. They are 300 times sweeter than sugar and apparently nontoxic. Chromatography was carried out by applying a mixture of steviosides obtained by extraction of dried, ground leaves with an aqueous 85 % methanol solution to Chromarods S or S II[257]. The rods were developed for 1.5 h in a chloroform-methanol-water system (62:28:10, lower phase). The mixture yielded five peaks, three of which were identified as monoglucosylstevioside (R_F 0.19), stevioside (0.38) and steviolbioside (0.59).

stevioside
X = O, Y = glc

steviolbioside
X = O, Y = H

monoglucosylstevioside
X = glc, Y = glc

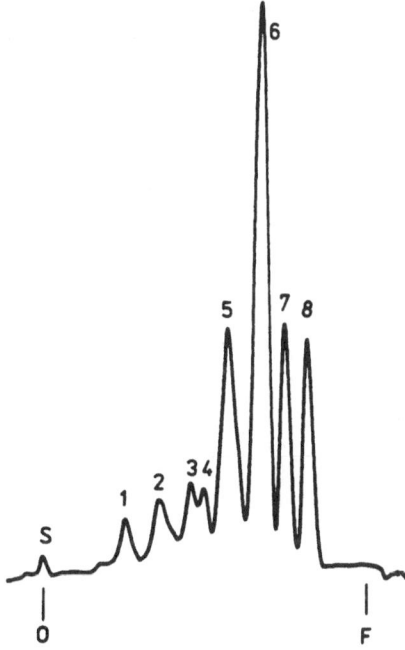

Esterification of the free hydroxyls in the saccharides leads to a decrase in the width of the individual separated zones and an improvement in the reproducibility of the analysis. An example is the TLC-FID of a mixture of octaacetylsacchararose and its partially deacylated derivatives on Chromarods S II in a benzene-*n*-hexane-methanol system (62 : 19 : 19). The chromatogram in Figure 36 depicts the sharp separation of mono-to octaacetylsaccharose, reflected in the low coefficients of variation of the determination (about 3 %)[258].

Fig. 36. TLC-FID of a mixture of various acetylated saccharoses on Chromarods S II. S saccharose; (1) monoacetylsaccharose, (2) diacetylsaccharose, ... (8) octaacetylsaccharose[258]

TLC-FID has also been used to study the partial benzoylation of 1,5-anhydro-4,6-O-benzylidene-D-glucitol by benzoyl chloride at −10 °C in pyridine. Chromatography with a benzene-ethylacetate (9 : 1) system separated the reaction products into four components: 1,5-anhydro-4, 6-O-benzylideneglucitol

(R_F 0.05), 1,5-anhydro-3-O-benzoyl-4,6-O-benzylideneglucitol (0.31), 1,5-an-hydro-2-O-benzoyl-4,6-O-benzylideneglucitol (0.44) and finally 1,5-anhydro-2,3-di-O-benzoyl-4,6-O-benzylideneglucitol (0.44)[259, 260]. The two positional isomers (the 3-O-benzoyl and 2-O-benzoyl derivatives) were separated very well. These authors also used the TLC-FID method for studying the partial tosyla-tion of 1,5-anhydroxylitol (using Chromarods S II and a toluene-acetone sys-tem (9:1)). Further details of the separation are not given in the paper[260].

Chromarods S II can be used for satisfactory separation of the products of partial acetylation of acetamidoglucosides[261, 262] using a benzene-methanol-ethyl acetate-acetic acid system. It is interesting that, while isomers with gluco and altro configurations (peaks 2 and 3, Figure 37) are separated, another pair of isomers with the acetyl group in the 2- or 3-position form a single peak (signals 4 and 5).

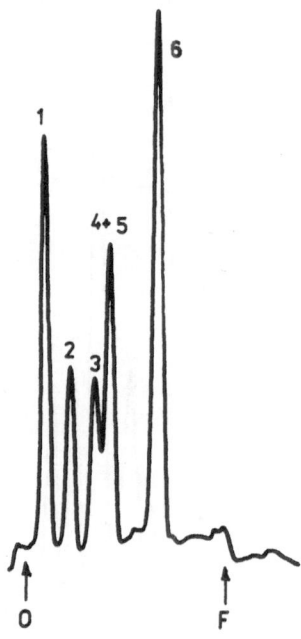

Fig. 37. TLC-FID of acetamidoglucosides on Chromarods S II using a benzene-ethylacetate-metha-nol-acetic acid system (40:5:5:1)[261]. (1) methyl-4-amino-4,6-dideoxy-α-D-glucopyranoside; (2) methyl-4-acetamido-4, 6-dideoxy-α-D-glucopyranoside; (3) methyl-4-acetamido-4, 6-dideoxy-α-D--altropyranoside; (4) methyl-4-acetamido-4,6-dideoxy-3-O-acetyl-α-D-glucopyranoside; (5) methyl--4-acetamido-4, 6-dideoxy-2-O-acetyl-α-D-glucopyranoside; (6) methyl-4-acetamido-4, 6-dideoxy---2, 3-O-diacetyl-α-D-glucopyranoside

This section should also include data on the TLC-FID of saccharose esters (and of a number of alcoholic sugars) with higher fatty acids, but as these substances are typical surface-active compounds, they will be considered in the section on surfactants.

2.2.4 Food Additives

The application of TLC-FID to food chemistry is limited to a few papers on the analysis of artificial flavours, preservatives and antioxidants and some special additives and foreign substances.

Kaneshima et al. published one of the first works in this field in 1974[263], dealing with the determination of the dialkyl esters of phthalic acid. These are plasticizers that can contaminate food stored in plastics. A hexane extract of the contaminated foods was applied to Chromarods S and eluted in an n-hexane--diethyl ether system (4:1) and detected in the FID.

Bindler, et al.[210] published an extensive paper on the capabilities and limitations of TLC-FID in the determination of artificial flavours of the phenolaldehyde type in sugar, antioxidants in dried potatoes and benzoic acid in beverages. They extracted phenolaldehydes from vanilla sugar (5 g) using a mixture of diethyl ether and pentane (7:3) or 95% ethanol. They then added an internal standard (veratrine aldehyde or diethyl phthalate) and concentrated the solution to a volume of 10 ml. An amount of 1 to 2 µl of the extract was applied to Chromarods S and the individual components were separated in a benzene--methanol system. The reproducibility of the determination, found from 15 measurements, was quite good. The responses for the individual aldehydes in the FID were comparable, but better than the corresponding response for diethyl phthalate (Table 41).

Table 41 TLC–FID of Phenolaldehydes and Antioxidants (Chromarods S)[210]

No.	System	Vol. %		Component				
				p-Hydro-xyben-zaldehyde	Vanillin	Veratrine aldehyde	Anisalde-hyde	Dietyl-phthalate
1	benzene methanol	99 1	R_F C. V. % K_{FID}^a K_{FID}^b	0.26 2.9 0.93 1.33	0.41 5.5 1.10 1.48	0.54 3.2 1.00 –	0.70 10.4 – –	0.53 – – 1.00
				BHT		di-BHA		BHA
2	diethyl-ether n-hexane	2.5 97.5	R_F	0.24		0.57		0.75

K_{FID} area of the signal for the internal standard (a veratrine aldehyde, b diethylphthalate) divided by the area of the signal for the same mass amount of the analysed substance.

The determination of benzoic acid in fruit juices was also successful. Diethyl ether extracts containing an internal standard, 3, 5-di-tert-butyl-4-hydroxyanisol (di-BHA) were applied to the rods. The rods were developed in an n-hexane, acetic acid system (99.2 : 0.8) and detected in the FID. The response for benzoic acid was 25 % lower than that for di-BHA (R_F, benzoic acid 0.32, di-BHA 0.62)[210].

In contrast, an attempt to determine antioxidants such as butylhydroxyanisole (BHA) and butylhydroxytoluene (BHT) was unsuccessful. These substances had to be extracted using alcohol, which also extracted other components that interfered in the separation and scanning in the FID. A model mixture of these substances can be separated using a diethyl ether-hexane system (Table 41). It is possible that better results would be obtained by multiple development of the extract in solvent systems with increasing polarity. The three-step elution of the components of commercial chewing gum is an example of successful application of this method. The first elution is carried out to a distance of 12 cm from the start using n-hexane and chromatographs the least polar waxes (R_F 0.85). The second elution with a mixture of benzene, chloroform and formic acid (90 : 8 : 2) to a distance of 10 cm separated resins (R_F 0.66 and 0.46) and the gum bases (0.34 and 0.18). Final development with a chloroform-methanol system (96 : 4) to a distance of 2 cm separates monoacylglycerol (R_F 0.08) from the remaining polar substances at the start[175].

2.2.5 Cosmetics

Of the aromatic substances used in cosmetics and in perfuming tobacco, a mixture of cumarine (2-benzopyrone) and cumaric acids (hydroxycinnamic acids) were separated using a chloroform-methanol-formic acid system (95.6 : 3.8 : 0.6). The chromatogram consists of three clearly defined signals. The closest to the start consists of a mixture of m- and p-cumaric acids (R_F 0.35), followed by o-cumaric acid (0.55) and finally by cumarine (0.77)[264].

In cosmetics research, the TLC-FID method is used primarily in series analyses and in analysis of competitors' products such as cosmetic creams, lipsticks, etc. Cosmetics usually contain ten or more components and thus these substances should first be distributed between water and chloroform[265].

Creams are extracted with chloroform and both separated phases, the aqueous and the chloroform, are dried. Chromatography on Chromarods S or S II using system no. 1 (Table 42) can be used to determine the glycerol and propylene glycol contents in the water-soluble fraction. The chloroform fraction is analysed on the same rods using systems nos. 2 and 3. The latter is used for separating low-polarity components[265].

In a more recent paper, good separation of all the polar and non-polar

Table 42 TLC–FID of Cosmetic Creams

(a) Solvent system (vol. %)

System no.	Benzene	Chloro-form	Acetic acid	Formic acid	Diethyl ether	n-Hexane	Methanol	Refer-ence
1		79.6		0.5			19.9	
2			1.0		9.9	89.1		265
3	20					80		
4 (a)	20					80		
(b)				0.5	15.0	84.5		36
(c)		79.7		0.3			20.0	
5 (a)					20	80		
(b)	25				75			265
(c)	40[a]	40					20	
(d)	35[a]	52.7		3.5	8.8			
6 (a)	20[a]					80[b]		
(b)					20	80		265
(c)	40[a]	40					20	
(d)	35[a]	52.7		3.5	8.8			

[a] toluene instead of benzene; [b] petroleum ether instead of n-hexane.

(b) R_F values

System no.	Chro-marods	G	PG[a]	MG	DG	ROH[b]	FA	TG	Lan[c]	PO[a] SQ[e]	W[f]
1	S II	0.45	0.62								
2[g]	S II	0.00	0.00	0.03	0.11	0.24	0.45	0.60	0.75	0.75	
3	S II					0.00	0.00	0.00	0.30	0.76	
4	S II	0.10		0.17	0.20	0.25	0.44	0.50	–	0.70	0.60[i]
5[h]	A				0.37		0.11				0.64[j]
6 (a, b)	A					0.23			0.31	0.61	0.90
										0.65	
										0.74	
6(c, d)	A		0.22	0.74							

[a] propyleneglycol; [b] fatty alcohol; [c] lanolin; [d] paraffin oil; [e] squalene; [f] wax; [g] plus tri-(2-methylhexyl)-glycerol R_F 0.53; [h] plus non-ionic surfactant R_F 0,26, silicon oil 0.88 and diesters of propyleneglycol with caprylic acid and capric acid R_F 0.47; [i] decylpalmitate; [j] isopropylmyristate.

components was attained without prior extraction by stepwise elution with three to four different systems, using both Chromarods S II[36] and A[209]. When chromatography is carried out using system no. 4 (Table 42), a benzene-n-hexane solution is used in elution to 10 cm to separate paraffin oil from decylpalmitate and then an n-hexane-diethyl ether-formic acid system is used on a 7 cm pathway to separate triacylglycerol, fatty acid (tails slightly) and fatty alcohol. Finally the most polar system (4(c)) is used to separate glycerol, monoacylglycerol and diacylglycerol (Figure 38)[36].

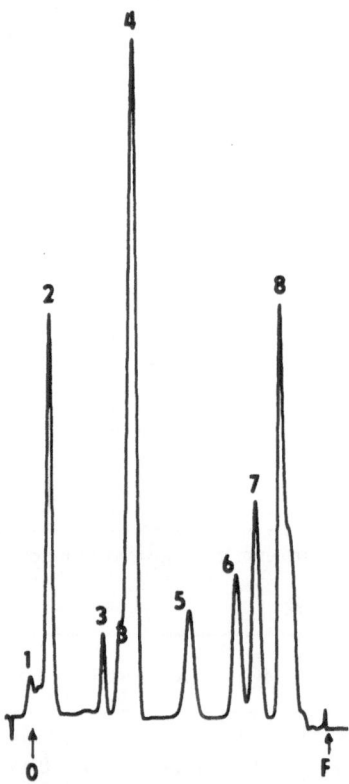

Fig. 38. TLC-FID of cosmetic cream on Chromarods S II using system no. 4 (Table 42). (1) glycerol; (2) monoacylglycerol; (3) diacylglycerols, 1,2 and 1,3 isomers; 4 fatty alcohol; (5) fatty acid; (6) triacylglycerol; (7) wax; (8) paraffin oil

Stepwise elution on Chromarods A using system no. 5 (5(a) to 12 cm, 5(b) to 6.5 cm, 5(c) to 4 cm and 5(d) also to 4 cm) can be used for good separation of silicon oil from isopropylmyristate; acylglycerol and non-ionic emulsifiers form a single peak between diacylglycerol and fatty acid. The polar substances remain unseparated at the origin; these include triethanolamine and some perfumes. System no. 6 is useful for the analysis of cosmetic creams by a combination of stepwise elution and elution with selective scanning. The first step involves stepwise elution with petroleum ether-toluene to a distance of 12 cm, and petroleum ether and diethyl ether to 10 cm. This procedure leads to separation of most of the non-polar and medium-polar substances. After selective scanning, the components remaining at the start are eluted by two systems based on chloroform and toluene (system 6(c) to 12 cm and system 6(d) to 4 cm), separating triethanolamine, propylene-glycol and monoacylglycerol[209].

2.3 Applications in the Chemical Industry and Related Fields

2.3.1 Surfactants and Detergents

Surfactants are surface-active substances containing at least one non-polar and one polar, hydrophilic group. The non-polar part of the molecule is usually an aliphatic or cycloaliphatic chain. The hydrophilic group generally consists of a carboxyl, sulphate, sulphonic or phosphate group (anionic surfactant), a quaternary ammonium group (cationic surfactant), amino group and carboxyl, aminoxide group (amphoteric surfactant) or a greater number of ether or hydroxyl groups (non-ionic surfactant). Detergents are substances that remove dirt (washing and cleaning agents) and contain surfactants as well as organic and inorganic builders and other substances with special effects.

The strongly polar group in the molecules of surface-active substances, their affinity for the surface of active adsorbents, tendency to associate in micelles in both aqueous and non-aqueous elution systems, and relatively small differences in the structures and volumes of the non-polar parts suggest a low probability of TLC-FID application in this technically important branch of industry. Nonetheless, several papers have dealt with the separation of these substances on Chromarods. Provided that substances with different numbers of hydrophilic groups or with very different polarity are to be separated, then good results can be obtained[209, 266—268].

A typical example is the separation of linear alkylbenzenesulphonates ($C_{10} - C_{15}$) from fatty acid soaps ($C_{16} - C_{22}$), which are the principal anion-active surfactants in low-foaming detergents used in automatic washers. An ethanol extract of the detergent is applied to the thin layer. The extract is first converted from the sodium salts to the free acids either by filtration through a column of a suitable cation exchanger or by acidification with dilute sulphuric acid[266]. The separation can be carried out using systems containing water[266—268]. When anhydrous systems are used (e. g. no. 1, Table 43), the signal of alkylbenzenesulphonic acid is quite broad and poorly separated from the signal of the fatty acid.

Although the dependence of the signal area for linear alkylbenzenesulphonic acid (LAS) and fatty acid on the amount applied is not very linear (Figure 39), the reproducibility of the determination using a calibration curve is quite good.

Table 43 TLC–FID of n-Alkylbenzenesulphonic (LAS) and Fatty Acid (FA) on Chromarods S
(a) Solvent system

System no.	Benzene	Chloroform	Acetone	Methanol	Water	Reference
1		90		10		266
2	85		5		10	266
3	50		10		40	266
4[a]		80		20		268

[a] separation not given

(b) R_F values

Component	System no.		
	1	2	3
LAS	0.43	0.13	0.24
FA	0.75	0.83	0.83

The response for alkylbenzenesulphonic acids in the FID is about 30 % lower than the corresponding response for fatty acids[266].

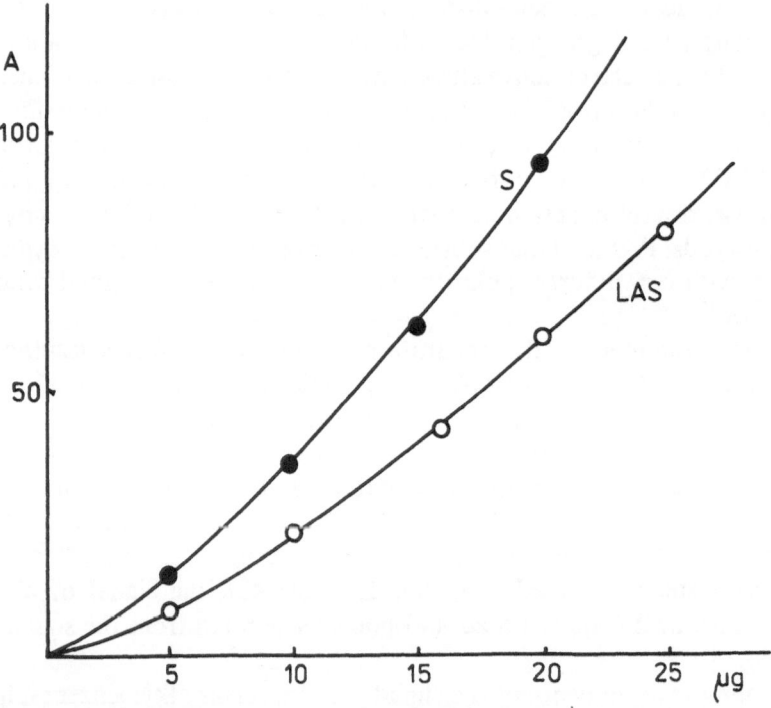

Fig. 39. Dependence of the response (A) for n-alkylbenzenesulphonic acid and soap on the applied amount. Chromarods S, system no. 2, Table 43[266]

System no. 4 is also useful for separation of binary mixtures of sodium alkylbenzenesulphonates and non-ionic adducts of ethylene oxide, e.g. of isononyl phenol with nine moles of ethylene oxide[268].

Sulphonated 2-alkenes (α-olephinsulphonates) are also industrially important surfactants. Considerable differences in polarity permit relatively simple chromatographic separation of alkene disulphonates from hydroxyalkane sulphonates and hydrocarbons using systems based on benzene, ethyl acetate and ethanol or chloroform, acetone and methanol (Table 44).

Table 44 TLC–FID of 2-Alkenesulphonates on Chromarods S

No.	System	Vol. %	Alkenedi-sulphon-ates	Hydroxy-alkane-sulphon-ates	Alkene-sulphon-ates	Hydro-carbons	Refe-rence
1	benzene	28.6					
	methyl ethyl ketone	28.6					
	ethanol	28.6	0.17	0.46	0.62	0.90	266
	water	9.4					
	14N ammonium hydroxide	4.8					
2	chloroform	59.3					
	acetone	14.8					
	n-hexane	3.7	0.14	0.40	0.52	–	269
	methanol	14.8					
	pyridine	7.4					

The sharpness of the separation of alkenesulphonates depends on proper activation of the rods. Optimal results can be obtained by soaking for at least 1 h in concentrated sulphuric and washing in distilled water prior to combustion in the FID[269].

Of the other anion-active surfactants, perfluoroactanesulphonic acid has been determined by the TLC-FID method using Chromarods S II and an ethyl acetate-methanol system (9 : 1), with an R_F value of about 0.50. The sensitivity limit is below $0.05\,\mu g$[270].

Mixtures of fatty amines and some cationic surfactants were also separated on Chromarods S II. These substances differed in the number of high-molecular weight substituents on the quaternary carbon (Table 45)[266].

TLC-FID has a wide application in determinating the distribution of the individual non-ionic ethylene oxide derivatives obtained by ethoxylation of fatty acids, fatty alcohols, alkylphenols, etc. A typical example is the chromatogram of the adduct of nine moles of ethylene oxide to isononyl phenol, yielding a mixture of at least twelve individual compounds[271]. Chromatograms of the adducts of isononylpolyglycol ethers with 2–17 moles of the initial ethylene

Table 45 TLC–FID of Fatty Amines and Quaternary Ammonium Salts on Chromarods S

(a) Solvent system

System no.	Benzene	Chloroform	Methanol	Methyl ethyl ketone	Ethanol	Formic acid	Water
1		93.5	1.5			5.0	
2	30			30	30		10
3		69.2	4.2				26.6

(b) R_F values

System no.	RNH$_2$	R$_2$NH	RB$_z$X	RX	Reference
1	0.13	0.58			180
2			0.15	0.66	266
3			0.50	0.68	180

R = C$_{16}$–C$_{18}$; RB$_z$X = RB$_z$(CH$_3$)$_2$ NCl; RX = R(CH$_3$)$_3$ NCl; B$_z$ = benzyl.

oxide[180, 209, 272], the adduct of oleic acid with six moles of ethylene oxide and six moles of propylene oxide[175] and the ethoxylated fatty acids[209] all yield detailed information. An elution system of ethyl acetate-acetone-water (74.5 : 21.3 : 4.2) was used in all cases.

Figure 40 depicts an example of the application of TLC-FID to the identification of commercial isononylphenolpolyglycol ethers with various degrees of

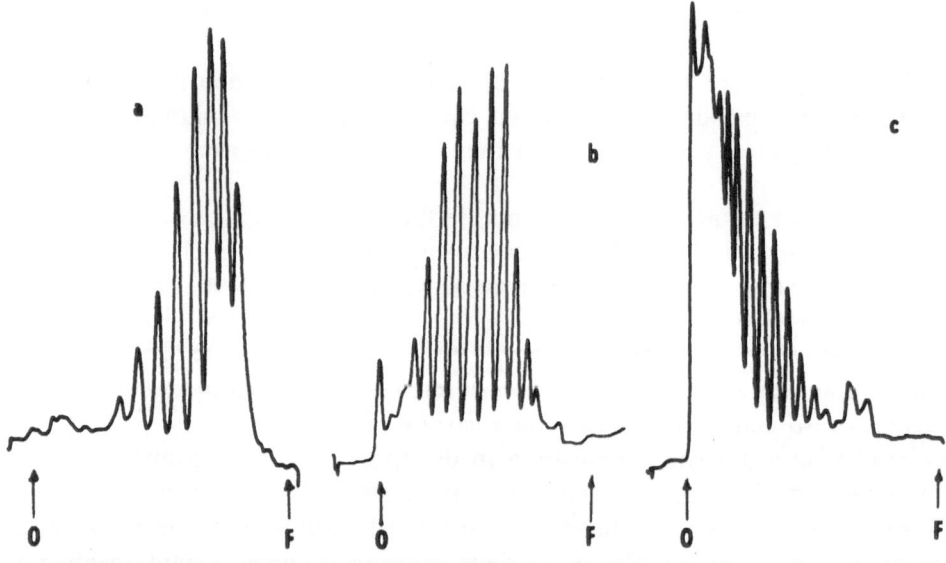

Fig. 40. Use of the TLC-FID technique for identification of the degree of ethoxylation (n = number of added moles of ethylene oxide (E O) per mole of alkylphenol) for isononylphenolpolyglycolethers. Chromarods S II, system given in the text. (a) 6 E O, (b) 9 E O, (c) 15 E O

ethoxylation. Each of the peaks in the chromatogram represents one individual alkylphenolpolyglycol ether. The constant difference in the R_F value for the individual adducts (ca. 0.6) is noteworthy, and corresponds to the lengthening of the polyglycol chain by a single ethylene oxide unit. The same difference was observed for the adducts of ethylene oxide to fatty alcohol[36].

The Iatroscan-Chromarod S technique using an ethyl acetate-ethanol-ammonia system (65:22.5:12.5) was used to determine the ratio between the anionic and non-ionic substances in sulphated alkylphenoxy ethoxylated surfactants. The results agree well with the analysis of the same samples by the HPLC method on Hypersil 5-CPS using an n-hexane-ethanol mobile phase (7:3)[273].

Chromatography on Chromarods S can be used to study the condensation of fatty acids (e. g. oleic acid) with monoethanolamine, diethanolamine and triethanolamine[209]. Fatty acid monoethanolamides yield primarily non-ionic amides ($RCONHC_2H_4OH$) and a small amount of ester ($RCOOC_2H_4NH_2$). The reaction with diethanolamine forms amides ($RCON(C_2H_4OH)_2$), esteramines ($NH(C_2H_4OCOR)_2$) and esteramides ($RCON(C_2H_4OCOR)_2$). Triethanolamine yields the monoester ($N(C_2H_4OH)_2C_2H_4OCOR$), diester ($N(C_2H_4OCOR)_2C_2H_4$-OH) and triester ($N(C_2H_4OCOR)_3$). The procedure used in the analysis of these compounds is apparent from Table 46.

Table 46 TLC-FID of Oleic Acid Ethanolamides on Chromarods A[209]

(a) Solvent system (vol. %)

System no.	Chloroform	Toluene	Methanol	Diethyl ether	Formic acid
1	49.3	49.2	1.5		
2	47.6	47.6	4.8		
3	52.6	35.1		8.8	3.5

(b) R_F values

(b)(i) Reaction of oleic acid with monoethanolamine. Elution procedure: system no. 1 to 12 cm, then systems nos. 2 and 3 to 5 cm.

Monoethanolamine	Oleic acid	Amides	Esters
0.00	0.28	0.26	0.84

(b)(ii) Reaction of oleic acid with diethanolamine.
Elution procedure: system no. 1 to 12 cm, system no. 2 to 5 cm and system no. 3 to 2.5 cm.

Diethanolamine	Oleic acid	Amides	Monoesters	Diesters	Esteramides
0.00	0.14	0.28	0.63	0.77	0.77

(b)(iii) Reaction of oleic acid with triethanolamine.
Elution procedure: as for (b)(i).

Triethanolamine	Monoesters	Diesters	Triesters
0.00	0.16	0.54	0.84

The separation of the non-ionic derivatives of sugars, such as the esters of fatty acids with saccharose and sorbitol, is dependent primarily on the number of acyl residues in the surfactant molecule. Because of the large number of possible positional isomers, the individual signals are broader than for the adducts of ethylene oxide. Chromatograms published by the Iatron company[175], obtained by elution in a chloroform-methanol-formic acid system (88.2 : 2.9 : 8.2) are only informative in nature and the individual signals are not identified.

Very good separation of the non-ionic esters of the fatty acids with pentaerythrite were obtained using a four-step elution on Chromarods A. In addition to these surfactants, free pentaerythrite and free fatty acid can also be determined simultaneously. This procedure can thus be used to follow the esterification process[211] (Table 47).

Table 47 TLC–FID of Non-ionic Esters of Fatty Acids with Pentaerythritol[211].

(a) Solvent systems (mass %)

System no.	Elution path cm	Chloro-form	Acetone	Diethyl ether	Formic acid	Metha-nol	Petro-leum ether	Toluene
1	12	–	15	–	–	–	85	–
2	8	–	–	9.8	–	11.8	78.4	–
3	4	35.1	–	8.8	3.5	–	–	52.6
4	2	40	–	–	–	20	–	40

(b) R_f values

Pentaerythrite	Fatty acid	Monoester	Diester	Triester	Tetraester
0.15	0.28	0.46	0.56	0.67	0.86

In conclusion, an original application of the Iatroscan instrument to study the effectiveness of detergents will be discussed[274]. Here, the surfactants are not determined, but rather the greasy dirt extracted from dirty and washed laundry. The authors used the same system as that employed for chromatography of lipids obtained from leather[143] (see Section 2.1.1).

Human sebum adsorbed on laundry is a complex mixture of lipids containing about 25 % fatty acids, 55 % triacylglycerols, 9 % waxes, 9 % hydrocarbons (primarily squalene), 3 % cholesterol and a small amount of phospholipids[275]. Washing in alkaline detergents involves saponification of the fatty acids and extraction into the wash water; however, acylglycerols mostly remain adsorbed on the fibres. Acylglycerols can be removed better by enzymatic hydrolysis through the action of lipase[276]. This process can be followed on Chromarods S II impregnated with boric acid[59] (see also Section 2.2.1).

2.3.2 The Petroleum Industry and Related Fields

TLC-FID finds wide application in the fuel oil industry and related areas. It is used, for example, in control of the composition of crude oil[277], heating and light oils and the products of the combustion of coal[278—281].

Kessler and Müller[282] used classical TLC in the late 1960s to separate polycyclic hydrocarbons. Ten years later, a work was published dealing with thin-layer chromatography with flame-ionization detection. Stepwise elution of petroleum from Kuwait using n-hexane and a mixture of n-hexane with benzene yielded signals for polar resins, aromatics, alkylaromatics, acyclic compounds and aliphatic hydrocarbons[283]. Later, a method was developed involving triple elution for the analysis of heating oils[271, 279, 280, 284, 285].

TLC-FID is much faster than column chromatography, which has been used up until now. In the latter method, the separated fractions must be dried and weighed and TLC control often indicates that these fractions are not pure[285]. The overall yield is rather low (85–95 %) and depends on the origin of the analysed material[285—287]. The column packing must be renewed after each analysis[284]. An attempt to replace column chromatography by the HPLC method failed because of problems involved in detection[285, 288].

Prior to TLC-FID analysis, the oil sample (b. p. > 250–300 °C) is freed of asphaltenes in a similar manner to the preparation for the analysis by column chromatography (the ASTM D 2007 method)[284]. For example, a sample of oily residues from the atmospheric distillation (b. p. > 390 °C, 30 g) is mixed with redistilled n-heptane (600 ml) and refluxed for 1 h. After cooling, the solution is filtered through a Whatman No. 1 filter paper and the filter paper is also extracted with n-heptane in a Soxhlet reflux apparatus. The combined filtrates are then evaporated in vacuo. In the analysis of liquid coal oil, the sample is freed of solids by filtration of a methylene chloride solution. The latter can also be purified as in the SARA method by precipitation with n-pentane. In another paper[279] the liquid fractions obtained by hydrogenation of various types of coal at temperatures above 400 °C are extracted with benzene and asphaltenes are precipitated by adding of n-hexane.

Asphaltenes are determined in bitumen, heavy oils and synthetic fuels in the presence of malthenes (a mixture of oils and resins) by dissolving the samples (200 mg) in benzene (25 ml) and separating the insoluble substances by filtration (Whatman no. 40). The asphaltenes remain in the filtrate[289].

Systems for the separation of the components of heating oils and lubricants are based on aliphatic and aromatic hydrocarbons (n-hexane, benzene, toluene) and, for the more polar components, contain small amounts of chlorinated hydrocarbons (dichloroethane, chloroform), lower alcohols (methanol, iso-propylalcohol) or ethyl acetate. A simple system (no. 1, Table 48) and single-step elution are sufficient only for the separation of groups of polar substances (e. g.

Table 48 TLC–FID of Derivatives of Oil and Coal

(a) Solvent system (vol. %)

System no.		Chroma-rods	Solvent system	Elution path cm	Reference
1		S II	n-hexane, isopropyl alcohol 95:5	8	289
2	(a)	S	n-hexane	10	283
	(b)		n-hexane, benzene 80:20	5	
3	(a)	S II	n-heptane	10	291
	(b)	A	chloroform, methanol 90:10	5	
4	(a)	S II	n-hexane	11	
	(b)	A	n-hexane, toluene 80:20	5	280
	(c)		dichlormethane, methanol 95:5	2	279
5	(a)		n-heptane	11	
	(b)	S	toluene	5.5	284
	(c)		chloroform, methanol 90:10	2	
6	(a)		n-hexane		
	(b)	S II	n-hexane, benzene 25:75	b	279
	(c)		benzene, methanol 95:5		
7	(a)		n-hexane	9	
	(b)	A	benzene[a]	5	285
	(c)		dichlormethan, methanol 60:40	2.5	
8	(a)		n-hexane	10	
	(b)	A	benzene[a]	5	278
	(c)		benzene, methanol 50:50	3	
9	(a)[c]		n-hexane	10	
	(b)	S II	n-hexane, benzene 90:10	4	290
	(c)		benzene, ethyl acetate 95:5	2.5	

[a] can be replaced by tetrahydrofuran; [b] not given; [c] after elution in system (a) selective scanning from 6 cm.

(b) R_F values

System no.	Resin		Aromatics	Saturated oils
	B	A		
1	a	0.67	0.67	0.67
2		0.14	0.44 – 0.59	0.81
4 (S II)	0.02	0.18	0.44	0.81
4 (A)	0.04	0.29	0.52; 0.62	0.71
5	0.08	0.23	0.44	0.73
6	0.05	0.26	0.44	0.73
8	+	0.26	0.42	0.80
9	0.00	–	0.36 – 0.44[b]	0.63

[a] asphaltenes R_F 0.00; [b] plus polyaromatics R_F 0.10.
Note: The R_F values are not given in refs. 291 (system 3) and 285 (system 7).

asphaltenes) and nonpolar components (malthenes). A simple elution system (benzene-cyclohexane, 5:95) was also used to separate anthracene (R_F 0.40) from n-docosane (0.80)[292]. More effective separation can be achieved by elution with two or three systems with increasing polarity.

The first system moves the hydrocarbons almost to the front; aromatics from a broad peak extending from the start to half-way along the rod. Polar substances remain at the origin (Figure 41). The next elution system, based on toluene or benzene, forms a single peak for the aromatics; this peak is split on Chromarods A. The part closer to the front is assigned to low-molecular-weight simple aromatics, and the other part to polyaromatics or neutral heteroaromatic compounds (system no. 4, rod A)[279]. The third system separates materials at the origin into two signals, corresponding to more polar resin B (which can also contain part of the asphaltenes)[284] and resin A.

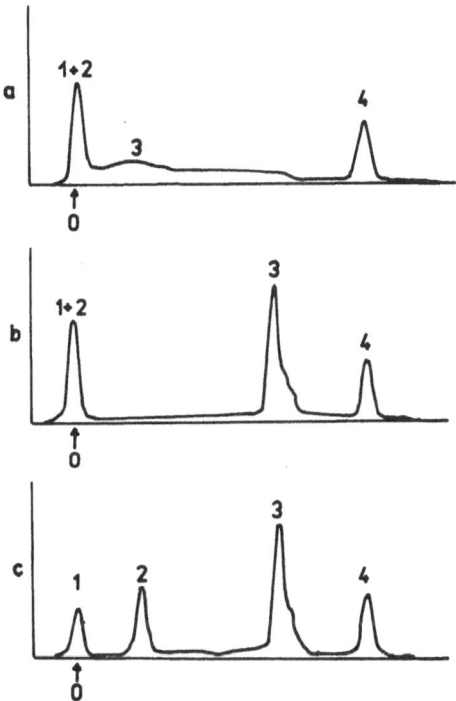

Fig. 41. TLC-FID of residue from the atmospheric distillation of crude oil. Chromarods S, system no. 4, Table 48. (a) elution with system 4(a) to 11 cm; (b) system 4(a) to 11 cm + 4(b) to 7.5 cm;(c) systems 4(a) to 11 cm, 4(b) to 7,5 cm, 4(c) to 2 cm. (1) asphaltenes and polar resins; (2) resin; (3) aromatics; (4) saturated hydrocarbons[280]

In the analysis of fractions obtained by column chromatography[293] of refined heavy distillates of Soviet petroleum on Chromarods A, the aromatics separated into three fractions, corresponding to monoaromatic, diaromatic and polyaromatic compounds[36]. The individual fractions were not chromatographically pure. For example, the second fraction contained saturated hydrocarbons in addition to more polar oxidized products; the peaks of these substances on the chromatogram corresponded to resins A and B (Figure 42). Polar nitrogen--containing and sulphur-containing substances have the same R_F value as resin A.

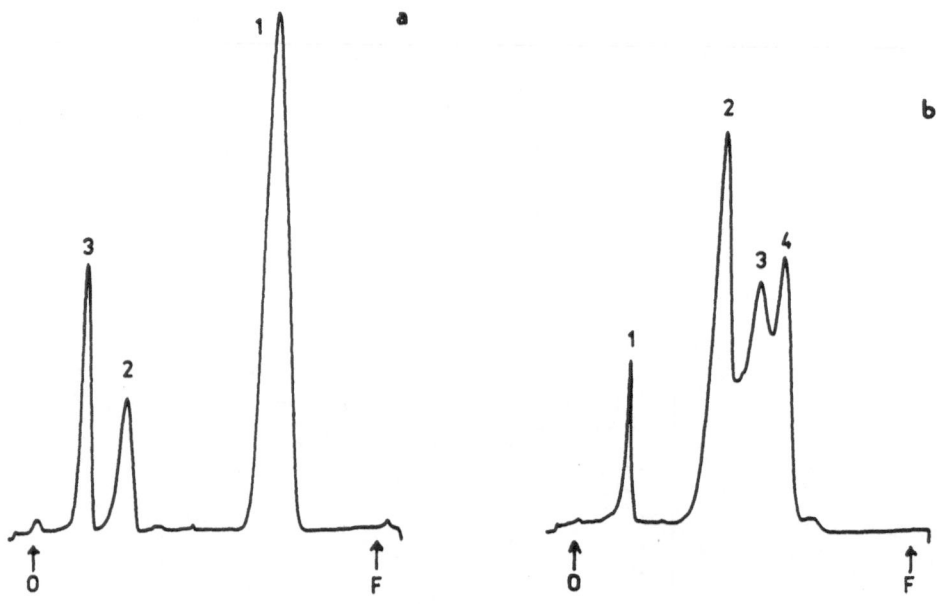

Fig. 42. TLC-FID of fractions of refined heavy distillate of Soviet crude oil, obtained from column chromatography[36]. Chromarods A, system no. 5, Table 48. (a) second fraction: (1) saturated hydrocarbons, (2), (3) oxidized products; (b) fourth fraction: (1) sulphur-and nitrogen-containing substances; (2) polyaromatics; (3) diaromatics; (4) monoaromatics

The use of the Iatroscan technique to study the oxidation of aromatics is also discussed in the paper of Ray et al.[284].

It is recommended that samples with high resin contents be applied in smaller amounts, about one half to one quarter of the usual volume. In the chromatography of oils obtained by coal liquefaction (with low saturated hydrocarbon contents), it is recommended that elution with n-heptane be followed by pyrolysis of the upper part of the rod and that the unpyrolysed residue be separated by stepwise elution with three systems-a 15 % solution of toluene in n-heptane to 10 cm, chloroform to 6 cm and a 5 % solution of ether in chloroform to 2 cm. The chromatogram consists of four well-separated peaks corresponding to two polar resins and two groups of aromatics[284].

In a study of the relation between the fuel properties and the exhaust emission, Obuchi et al.[294] used TLC-FID technique to the analysis of an organic extract of diesel exhaust particulates. Prior to use, activated rods were put in a chamber with the 58 % relative humidity. Typically, 2 µl of dichloromethane extract of exhaust were spotted on Chromarods S II and the rods were developed in n-hexane to 10 cm from the origin. The R_F values of aromatics ranged from 0.11 for 1,2-benzopyrene to 0.50 for dodecylbenzene. Normal alkanes exibited almost the same R_F value (0.70), regardless of the length of the carbon chain. Polar compounds, having at least one heteroatom, remained at the start point. In addition, it is possible to separate the polar substances futher by using toluene as an elution solvent. In this case the R_F value of carbazole is 0,04, that of anthraquinone is 0.21, while the R_F value of higher polar compounds, such as acids and bases, are still almost zero. The detection limit is ca. 0.03 µg.

A procedure with selective pyrolysis of saturated hydrocarbons after the first elution with n-hexane was also used to improve the separation of polynuclear and a mixture of mono- and dinuclear aromatics in heavy oils and synthetic fuels (system no. 9, Table 48). The ethyl acetate content in benzene (system no. 9(c)) ensures good separation of the polynuclear aromatics from polar substances (resins from asphaltenes)[290].

Polynuclear aromatics, thought to be carcinogenic, form the main fraction of the coal tar pitch volatiles contained in the industrial atmosphere where large amounts of coal or coke are burned. The content of these substances in the air is generally determined gravimetrically. The contaminated air is filtered and the filter is extracted with benzene. The latter is evaporated and the residue is weighed. Up to 200 samples are analysed in this way at a single plant each day. Analysis by the TLC-FID method is simpler and yields more information. The extract from the filter is applied directly to the Chromarods and eluted in an n-heptane-isooctane system (1 : 1). The more polar aromatics remain close to the start and the relatively harmless oil migrates to the front[291]. A modification has been developed for application of this method to the control of the purity of air in aluminium smelters. This procedure permits separation of polynuclear aromatics (R_F 0.00 and 0.38), from harmless components, i. e. vegetable oils (0.20) and mineral oil (0.56) using a 5 % diethyl ether solution in n-hexane[287]. The agreement of the TLC-FID results with the official gravimetric method is very good (correlation coefficient 0.990)[287].

Some authors have stated that the dependence of the response on the concentration is linear[285, 289], while others have found an S-shaped curve, with poorer reproducibility[284]. This can be considerably improved by using a suspended developing system (see Section 1.2.3)[284]. The coefficient of variation depends on the types of components separated and on their contents in the analysed mixture; for example, for a saturated compound content of about 40 %, the coefficient of variation is about 7 %; at the same content of aromatics, it may be as much as 16 %[284]. Another paper dealing with the chromatography of anthracene oils gives a coefficient of variation of only 2 %[285]. A value of 2–4 % was found in the analysis of oils using system no. 9[290].

2.3.3 The Industrial Production of Technical Emulsions for the Metal, Textile and Leather Industries

Recently, analytical chemists have also considered the analysis of industrial emulsions. The main components of these substances are paraffin oil, acylglycerols, waxes, the methyl esters of fatty acids, anticorrosion and disinfectant additives, glycols, ethanolamines, surfactants, etc. These mixtures are used as lubricants and softeners in the textile and leather industries and also in the metallurgical industry for milling, pressing, rolling and machining of metals. Some of these substances, such as machining emulsion, are recycled; this leads to a gradual change in the composition, through chemical, biological or thermal degradation, or through adsorption. Poré et al. have published a number of papers on the development of a TLC-FID method. In spite of the complexity of the analyzed mixture, very good component separation can be attained with excellent reproducibility; the coefficient of variation for most components is less than 0.5 %[295—298]. The authors carried out a number of experiments demonstrating the advantages of multiple elution of a multicomponent mixture with varying polarity. They depict the TLC-FID chromatograms of products for the fat-liquoring process[295,297,299], emulsions and oils for milling aluminium and steel[296,300], etc. They employed four-fold stepwise elution to 12, 6, 4 and 2 cm using solvent systems with increasing polarity, such as n-hexane-chloroform (95:5), petroleum ether-diethyl ether (80:20), chloroform-toluene-methanol (2:2:1) and chloroform-benzene-formic acid (30:30:1).

For example, this system was used to separate products for degreasing of leather into aliphatic hydrocarbons (R_F 0.75), chlorinated hydrocarbons and waxes (0.62), triacylglycerols (0.42), mono- and diacylglycerols (0.36), fatty acids (0.17) and polar components (0.00)[299]. It is interesting to note that, except for the paper by Poré[299], the compositions of the solvent systems are not given in the other works cited. The authors recommend Chromarods A which apparently have better separation ability, longer lifetime and lower tendency to adsorb impurities from the air. The accuracy and reproducibility of the analysis is maintained by using three calibration mixtures simultaneously, containing all the components present in the analysed sample. To a certain degree, this eliminates errors resulting from the dependence of the correction factors for the FID on concentration[295—297,300]. The results of these papers are collected in a separate paper by Newman[301], which also gives the TLC-FID chromatogram of cutting oils obtained on Chromarods S II by simple elution using a chloroform-methanol-ammonium hydroxide system (70:28:2). The chromatogram contains a clearly separated double peak for paraffin oil (R_F 0.91) and non-ionic surfactants (0.89) and single peaks for cationic disinfactants (0.80), corrosion inhibitors (0.62), fatty amine (0.34), secondary amine (0.12) and alkylphosphate (0.00).

2.3.4 The Polymer Industry

Classical thin-layer chromatography has long been used in polymer research. Reviews of the separation of polymers by TLC methods were published in 1977[302–304].

TLC-FID is suitable primarily for the separation of low-molecular weight polymers and polycondensates, differing either in the number and type of bonded functional group, or in the content and structure of the original monomer. The first TLC-FID paper in this field deals with the separation of the components of liquid rubber (molecular weight 1200 to 2000), used industrially as a monofunctional adhesive[305, 306]. This involves chromatography of a mixture of telechelic prepolymers, 1, 2-polybutadiene and α, ω-bifunctional polybutadiene with chain termination by hydroxyl or carboxyl groups, prepared by reaction of the corresponding homopolymers with ethylene oxide or with carbon dioxide. An amount of 2 µl of the polymer mixture is applied to Chromarods S and separated by double elution in p-xylene (to 10 cm) and chloroform-tetrahydrofuran (90.9 : 9.1) (to 5 cm) systems, yielding zones corresponding to nonfunctional (R_F 0.90), monofunctional (0.50) and bifunctional (0.26) polymers. The type of bonded functional group does not appear to affect the R_F value. It is also interesting that the FID response is not dependent on the R_F value, as happens, for example in densitometry, where the response for a given substance close to the start is less than that at the front. The phenomenon is a result of diffusion from the inner part of the layer to the surface during migration.

Synthetic resins are mostly analysed by column chromatography with subsequent determination of the molecular weights and number of carboxyl and hydroxyl groups in the individual fractions. The application of TLC-FID yields information on the compositions of the fractions in a very fast manner. A solution of the resin in chloroform or tetrachloromethane (ca. 5 %) is applied to Chromarods S II and developed stepwise with chloroform-methanol-acetic acid (50 : 50 : 1) to 10 cm, (60 : 40 : 1) to 7 cm and (80 : 20 : 1) to 3 cm. The rod is dried for 3 min at 120 °C between the individual elutions and prior to detection. The chromatogram consists of four quite sharp peaks (R_F 0.00, 0.25, 0.55 and 0.80); the ratio of their areas is characteristic for the given resin. Separation of acrylated resins for paints on Chromarods A using an ethyl acetate-formic acid system seems to be possible, but the quality of the chromatogram does not ensure sufficient reproducibility for qualitative determination[307]. It would be preferable to carry out TLC-FID of a mixture of epoxide resins based on tetraglycidylmethylenedianiline with diaminodiphenylsulphone as a hardener; these are common components of composite materials with a resin matrix, used in building aeroplanes and space vehicles. A mixture of these two substances was separated using a benzene-tetrahydrofuran system (9 : 1). Epoxy resin yielded two signals, one close to the start and the second at the front. Diaminodiphenylsulphone yielded a single sharp peak with R_F ca. 0.6; the response depended linearly on the concentration of this component[308].

In a bulletin on the TLC-FID of the copolymer of styrene with methyl-methacrylate and butylethylacrylate by step-wise elution with diethyl ether and acetone, the Iatron company points out the necessity of standardization of the active layer in an evacuated dessicator and even presaturation in the elution tank[309] (see also Section 1.2.3).

The chromatography of the copolymer of styrene and acrylonitrile with an average molecular weight of $16–78 \times 10^4$ can be carried out using gradient elution with a tetrachloroethane-tetrahydrofuran system. An amount of 1 µg of the copolymer is applied to Chromarods S and the rods are placed in a chamber containing 70 ml of tetrachlorethane. Then four additions of 8.5 ml of tetrahydrofuran are made when the front of the solvent system reaches 0, 2.5, 5 and 7,5 cm front the start. The eluted rod is dried for 30 min at 120 °C prior to scanning in the FID. The analysed samples separate according to the styrene content in the polycondensate. The R_F value is directly proportional to the styrene content. Chromatography of the initial homopolymers demonstrated that the FID response does not depend on their chemical composition[306, 309].

The same method was utilized in research on the emulsion polymerization of polybutylacrylate and styrene[310].

The chromatographic analysis of the copolymers of styrene and cellulose is quite simple. These copolymers are prepared by the reaction of a solution of styrene in tetrachloromethane with viscose. The sample is first hydrolysed and then separated on Chromarods S in benzene[311] or toluene[180] to yield two components with R_F 0.02 and 0.76.

Analysis of polyepichlorohydrine by double elution in a dichloromethane-chloroform-hydrochloric acid system (90.6 : 9.0 : 0.4) on Chromarods S II yields three signals with R_F 0.87, 0.41 and 0.24[175].

The TLC-FID method has also been applied for the separation of a mixture of stabilizers used in the production of vulcanized rubber. Chromatography of Chromarods S II using a cyclohexane-toluene-acetone system (89 : 10 : 1) yields good separation of 2, 6-di-tert-amylhydroquinone (R_F 0.20) from the butylated reaction product of p-cresol and dicyclopentadiene (0.33), tris(nonylphenol)-phosphite (0.72) and complex compounds with mercaptan (0.05)[312].

2.3.5 Pesticides and Growth Regulators

The first publication dealing with the application of TLC-FID in this extensive branch dealt with the separation of O, O-dimethyl-O-(3-methyl-4-nitrophenyl)-phosphorthioate (Metathion) from 2-(methylpropyl) phenylmethylcarbamate (BPMC) in a mixture of insecticide emulsion concentrates using benzene (R_F

0.64 and 0.16) or benzene and methylethylketone (50 : 2.5, R_F 0.70 and 0.32). Separation was also carried out on a mixture of the product Baycid (O, O-methyl-O-(4-methylthio-3-methylthiophenyl)phosphorthioate) and Diazinon (O,O-diethyl-O-(2-isopropyl-6-methylpyrimidine-4-yl)phosphorthioate using a cyclohexane-chloroform-acetone-system (40 : 5 : 1, R_F 0.30 and 0.43). Both analyses were carried out on Chromarods S^{313}.

Since 1980, this method has gained importance in the synthesis and control of some pesticides and growth regulators. Chloromethanesulphonamides have been determined chromatographically after conversion to the corresponding amines by alkaline hydrolysis (using an n-hexane-benzene-methanol system, 79 : 20 : 1 and Chromarods S II)[52]. The N-alkylation of chloromethanesulphonanilides by suitable alkyl esters of chloracetic acid[314] has been studied. Samples of the reactants and products in the preparation of the herbicide Pyrazon have also been analysed[52]. At present, TLC-FID is the best method for analysis of growth regulators for grains; the main component of these substances is 4, 4-dimethylmorpholin chloride, in the presence of a small amount of 4-methylmorpholine hydrochloride[57], (Tables 14 and 49).

Table 49 TLC-FID of Some Pesticides

(a) Solvent system (vol. %)

System no.	n-Hexane	Benzene	Methanol	Chloro-form	Acetone	Diethyl ether	Refer-ence
1	79	20	1				314
2	19			80	1		52
3			8		62	30	56

(b) R_F values

System no.	
1	CMSA 0.16, Et-CMSA 0.35, ES 0.81
2	PCA 0.20 iso-PCA 0.55, PCC 0.74, TG 0.85
3	RW_1 0.34, RW_3 0.45, TAK 0.75

CMSA chloromethanesulphonanilide; Et-CMSA N(ethoxycarboxymethyl)chloromethanesulphonanilide; ES ethylstearate (internal standard); PCA, iso-PCA, PCC – see Table 11; TAK triacontane (internal standard); RW_1, RW_3 see Table 14.

Because of the relatively small carbon content, the responses of most pesticides in the FID are smaller than those for hydrocarbons and most lipids. The response related to that for an internal standard (triacontane, ethylstearate, triacylglycerol) depends on the detector conditions (see Tables 11 and 13 and the discussion in Section 1.2.3).

When studying the progress of the synthesis of pesticides and product purity, it is often necessary to determine substances present in the sample in amounts of less than 2 % by weight. It is then useful to apply much larger amounts of sample than normally. For example, when three or four-component mixtures are to be analysed with a sufficiently large difference in the R_F values, the applied amount (about 10 µg) can be increased by up to 10-fold[314]. This approach yields better reproducibility for minor components with a low response, such as 1-phenyl-4, 5-dichloropyridazone and chloromethanesulphonanilide (Table 50).

Table 50 The Effect of the Amount Applied on the Reproducibility of the Determination of the Pesticide Pyrazon[52] and the Reaction Components in the Alkylation of Chloromethanesulphonanilide with Chloroacetic acid[314].

(a) Analysis of Pyrazon. 15–75 µg of sample applied on Chromarods S, developed with system no. 2, Table 49. Scanning speed 0.42 cm s^{-1}, hydrogen flow-rate 180 ml min^{-1}, air 2.1 l min^{-1}.

PCC		iso-PCA		PCA	
µg	C. V. %	µg	C. V. %	µg	C. V. %
0.6	46.0	0.6	46.6	0.6	30.6
1.0	15.4	2.0	11.5	2.0	6.3
2.1	8.2	5.0	6.1	5.5	3.3
4.0	4.7	9.7	5.9	9.7	1.7
9.0	3.3	20.0	5.0	20.0	2.6
20.0	1.6	41.7	3.2	40.3	2.1

PCC = 1-phenyl-4,5-dichlorpyridazone (6);
iso-PCA = 1-phenyl-4-amino-6-chloropyridazone (6);
PCA = 1-phenyl-4-chloro-5-aminopyridazone (6).

(b) Analysis of the reaction components in the alkylation of CMSA. 15−100 µg of sample applied to Chromarods S II, system no. 1, Table 49. Scanning speed 0.31 cm s^{-1}, H$_2$ and air flow-rates as in Table 50 (a)

CMSA		Et-CMSA		E. S. (internal standard)	
µg	C V %	µg	C V %	µg	C V %
0.5	48.0	0.5	14.4	5.0	2.8
2.0	6.9	1.5	8.3	17.3	2.5
81.2	0.6	78.3	0.6	19.7	2.4

Because of its rapidity and consequent high productivity of analyses, TLC-FID has become an important method primarily for series control of the purity of mass produced pesticides, such as Pyrazon (also sold under the names Burex, Chlorazin, Pyramin and Phenazon). This substance was formerly analysed primarily by potentiometric[315], polarographic[316] and spectrophotometric[317] methods, later replaced by more effective separation procedures, especially GLC[318] and HPLC[52]. It follows from Table 51 that the results of the analyses of technical products by TLC-FID and HPLC are comparable and that the TLC-FID method reveals the presence of additional 'impurities' with an R_F value between those for PCA and iso-PCA (component 'X').

The complete separation of the components of Pyrazon by the TLC-FID method is apparent from the chromatographic trace (Figure 43) with very sharply separated symmetrical peaks and very low noise level.

Table 51 Comparison of the Results of Analysis of Samples of Pyrazon by the HPLC and TLC–FID Methods (mass %, average of four analyses)[52].

Sample no.	PCA		iso-PCA		PCC	
	TLC	HPLC	TLC	HPLC	TLC	HPLC
1	28.0	28.7	7.7	10.1	64.3	61.2
2	95.8	95.7	3.9	4.0	0.3	0.3
3	79.8	76.4	19.8	23.6	0.4	–
4	87.8	87.8	8.8	12.2	0.2[a]	–

[a] plus component „X", 3.2 %.

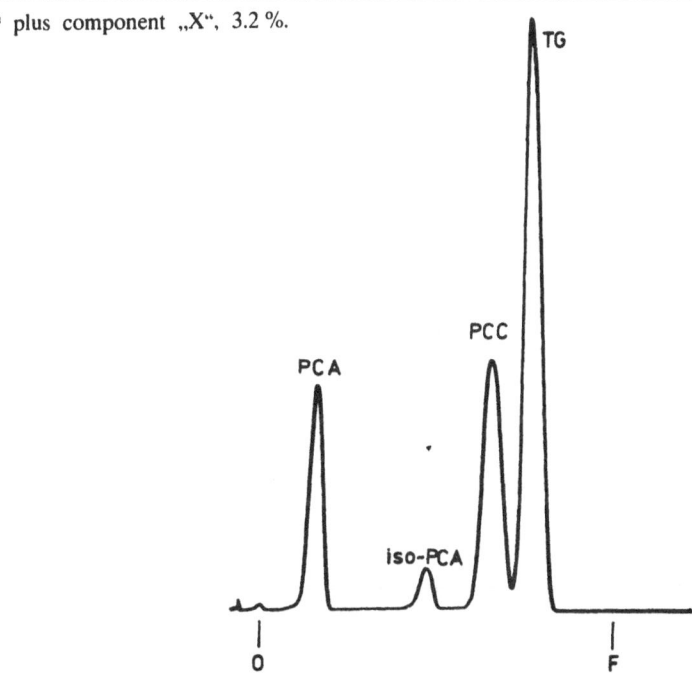

Fig. 43. TLC-FID of the components of the herbicide Pyrazone. Separation conditions given in Table 49 (system 2)[52]

The TLC-FID technique has been found to be more effective and less tedious than the GLC and HPLC techniques for separation of momilactones, acting as growth inhibitors, and phytoalexins, contained in the seeds of some plants, especially rice. The contents of both known types of momilactones in rice husks can be found by elution of the extract with a mixture of chloroform and ethylacetate on Chromarods S II. The TLC-FID analysis results can be used to estimate germination ability and resistance capacity of a given type of plant to fungus diseases[319].

Part 3 New Perspectives in TLC on Chromarods

Continuing development of the TLC-FID method has resulted in the improvement of the reproducibility and sensitivity. The introduction of new detecting
systems and the incorporation of micro-processing units offer futher flexibility
to the features of the existing techniques.

3.1 Increasing the Reproducibility and Sensitivity of the Determination

Factors affecting the reproducibility of TLC-FID measurements were discussed
in the last paragraph of Section 1.2.3. Of these factors, the greatest attention is
now being paid to even application of the product on the rod and improvement
of the geometry of the FID collector, as well as to the development of new thin
layers.

In normal application of the sample using a micropipette, a quite broad zone
is applied (up to several millimetres) and the distribution of the sample around
the circumference of the rod is not even. A new automatic applicator developed
by Read[320], employs the aerosol technique of application to a rotating rod. The
rods are placed in a holder and the sample application to a whole set of rods is
automated. In contrast to the usual method of application using a micropipette,
used so far, up to 20 µl of solution can be applied to each rod. The width of the
trace does not exceed a few tenths of a millimetre. This method yields much
better separation of multicomponent mixtures (the separated zones are narrower) and apparently considerably increases the reproducibility of the determination (the coefficient of variation decreases by up to five fold). Nowadays the
instrument is commercialy available.

The present detector design results in low collection efficiency for the ions
formed during ionization of the separated zones in the hydrogen flame. As the
outer mantle of the burner is grounded, it is probable that the lines of electrical
force beginning at the polarization centre of the burner will terminate on the
outer mantle without passing through the FID collector (Figure 44). The new

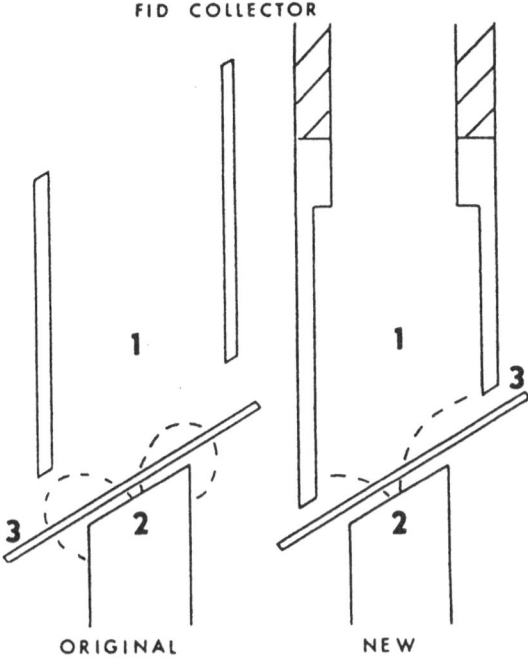

Fig. 44. Scheme of the original and new type of FID collector (1) collector; (2) burner; (3) Chromarod

type of collector is lowered to a position about 2.5 mm from the surface of the Chromarod rack, so that it is only 1 mm from the surface of the rods (compared to 6 mm for the original type). This improvement considerably increases the collection efficiency and sensitivity of the detector. In addition, the increased sensitivity is also reflected in improved linearity of the response at low concentrations.

The development of new thin layers has been studied and the processes which have been previously performed by manual control may, in future, be carried out by precise automatic means. This will enable a homogenous condition to exist in the silica particle wet suspension step thus avoiding sedimentation to occur and formation of a particle density gradient. The reproducibility of manufacture will allow and reflect better reproducibility during chromatographic analysis. Batch-to-batch and rod-to-rod reproducibility will also enable fewer rods to be used for averaging purposes, thus permitting a larger sample throughput.[324] Recent studies indicate that interrod differences can be further reduced by impregnating the Chromarods with a diluted solution of copper (II) sulphate[325-327].

3.2 New Detection Systems

Another detector, marketed since the middle of 1984, is the flame thermionic ionization detector (FTID)[321,322]. In contrast to the FID (see Section 1.2.3), ionization in the FTID occurs at a location far away from the chemically active hydrogen flame. Neutral molecules formed by pyrolysis of the sample in the hydrogen flame or by ion recombination impact on the surface of an electrically heated thermionic source. This source is coated with a ceramic layer impregnated with a caesium compound. When heated to a suitable temperature it is capable of emitting electrons which ionize electronegative components[321]. Thus organic compounds containing nitrogen or halogens are especially readily detected, while the ionization of hydrocarbon molecules is almost negligible.

The FTID developed for the Iatroscan system cannot be exchanged with the NP-type thermionic detector, originally popular in gas chromatography. In the FTID, the thermionic source is surrounded by a gaseous medium consisting of hot air, hot water vapour and combustion products from the analysed substance. In contrast, the NPD transducer is in a very hot, highly reactive gaseous boundary layer, formed of dissociated hydrogen and oxygen molecules and ions of the analyzed component developed in the reaction zone of the flame. In the

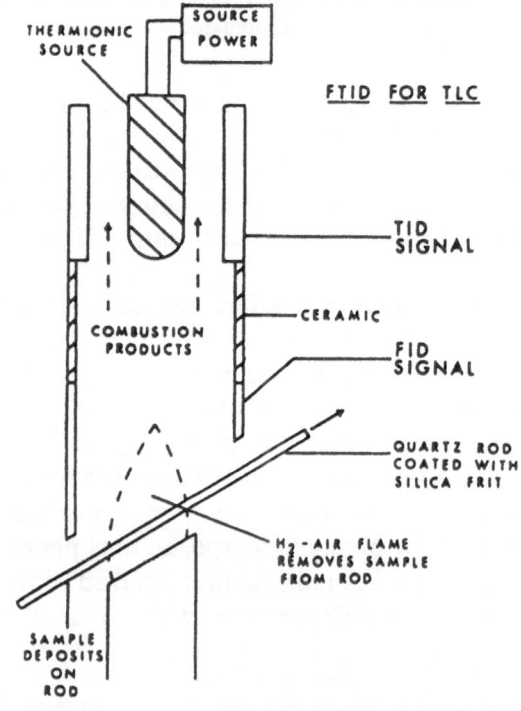

Fig. 45. Scheme of combined detection of Chromarods by the FTID-FID system

NPD, the degree of dissociation of H_2 and O_2 is maintained by the high surface temperature of the thermionic source (600–800 °C). In the FTID, the transducer operation temperature is usually lower (400–600 °C).

Installation of the new detector system in already existing Iatroscan instruments is relatively simple, consisting of exchange of the collector and lid replacement by a modified top cover attached to the FTID. The latter is connected to the FID collector via a ceramic tube (Figure 45). The combined detection system yields two signals for the pyrolysis of the separated (or unseparated) components-the universal FID signal and the specific thermionic signal (TID). The signal from the FID goes through the original amplifier resident in the Iatroscan. The TID signal passes through an additional electrometer module connected to the thermionic source power supply. The TID electrometer is a common GC-type with five amplification ranges (10^{-12} A mV^{-1} to 10^{-8} A mV^{-1}) and an attenuator ($1 \times$ to $1024 \times$). The use of a double-channel recorder or a suitable integrating computer permits recording of both signals (FID and TID) at once.

The operation in the FID-FTID mode provides a rapid means of quantifying the non-volatile organic matter content in solutions of a variety of origins, e. g. polluted water, alcoholic and non-alcoholic beverages etc. Liquid samples spotted on to the rod at five separate locations along each Chromarod can be detected (without chromatographic separation) to produce both total organic carbon and total nitrogen or halogen signals. The use of a ten rod array system means that fifty samples can be processed in the total scan time of 5–6 minutes[324].

A further detection system being prepared for application with the Iatroscan instrument is the flame emission photometric detector (EPD or FED) for analysis of substances containing sulphur and/or phosphorus[323].

3.3 New Model of the Iatroscan

In 1986 Iatron introduced a new version of the Iatroscan (Mk IV)[324, 328] (Figure 46). This instrument provides greater access to operating variables that can be controlled by micro-processor techniques which greatly extend the scanning capability and flexibility of the system. A feature of the model demonstrates the ability to select a control program to vary the scanning conditions, scanning path length as well as the scanning sequence. For example, a program can be selected to scan any single rod or any group of successive rods from the array. The number of blank-scans along the full length of the thin layer coating or of the area at the origin of any particular rod, can be arbitrarily fixed.

To facilitate the correct adjustment of the detector geometry relating to the rod surface, the hydrogen burner has been re-located. The cylindrical collector within the detecting system has been now electrically shielded.

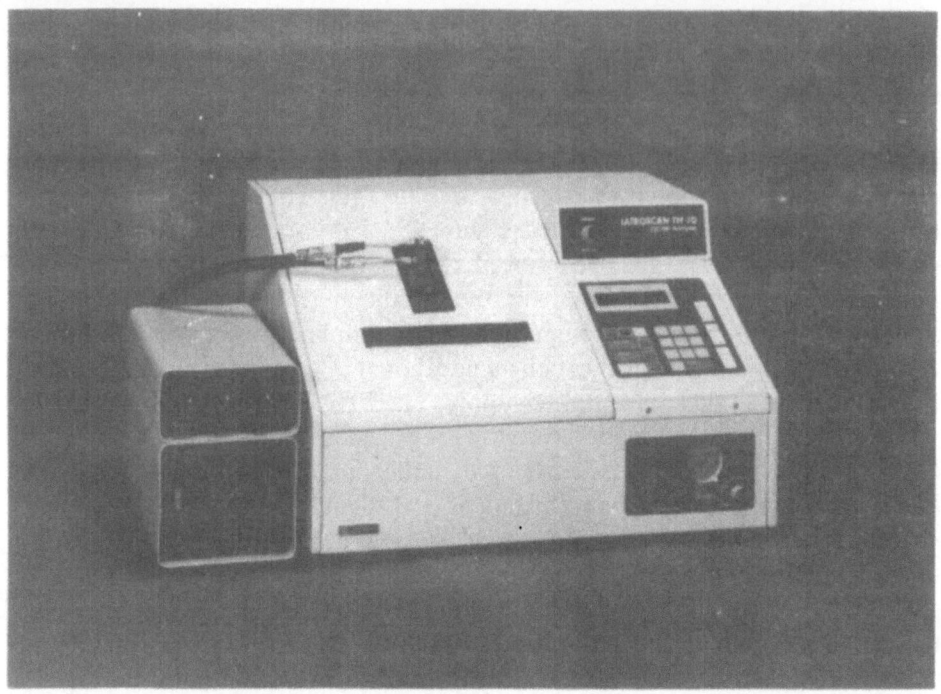

Fig. 46. The Iatroscan TH-10, Mark IV with the FTID and FID collectors, connected to the additional electrometer module and detector current supply

To improve the ergonomics of the system, all indicating and controlling units (including the air and hydrogen flowmeters) have been re-located at the front of the instrument. Other improvements are in the provision of a new external air supply, new current-type amplifier and stepping motors.

The collective modifications and improvements in the Iatroscan Mk IV system, in addition to the newly introduced detectors, are indicative that TLC--FID-FTID is a continuously developing method designed to keep abreast of changing technology within the field of modern chromatographic techniques.

References

1. STAHL, E.: Dünschicht-Chromatographie, Springer-Verlag, Berlin, 1962.
2. STAHL, E.: Thin Layer Chromatography. A Laboratory Handbook, Springer-Verlag, Berlin, 1969.
3. MACEK, K., HAIS, I. M., KOPECKÝ, J., GASPARIČ, J., RÁBEK, V., and CHURÁČEK, J. (eds.): Bibliography of Paper and Thin-Layer Chromatography 1973—1975 and Survey of Applications. Elsevier, Amsterdam, 1975.
4. DEJL, Z., MACEK, K., and JANÁK, J. (eds.): Liquid Column Chromatography, J. Chromatogr. Library, Vol. 3, Elsevier, Amsterdam, 1975.
5. TOUCHSTONE, J. C., and SHERMAN, J.: Densitometry in TLC Practice and Applications, Wiley, New York, 1979.
6. ZLATKIS, A., and KAISER, R. E. (eds.): High Performance Thin-Layer Chromatography, J. Chromatogr. Library, Vol. 9, Elsevier, Amsterdam, 1977.
7. KAISER, R. E. (ed.): Instrumental High Performance Thin-Layer Chromatography, Proceedings of the Second International Symposium, Institute for Chromatography, Bad Dürkheim, 1982.
8. TYIHAK, E., MINCSOVICS, E., and KÖRMENDI, F.: Hung. Sci. Instr. 55, 33 (1983).
9. KAUFMAN, H. P., and MUKHERJEE, K. D.: Fette, Seifen, Anstrichmittel 71, 11 (1969).
10. HAAHTI, E., VIHKO, R., JAAKONMÄKI, I., and EVANS, R. S.: J. Chromatogr. Sci. 8, 370 (1970).
11. MANGOLD, H. K., and MUKHERJEE, K. D.: J. Chromatogr. Sci. 8, 379 (1970).
12. MUKHERJEE, K. D., SPAANS, H., and HAAHTI, E.: J. Chromatogr. 61, 317 (1971).
13. PADLEY, F. B.: J. Chromatogr. 39, 37 (1969)
14. SZAKASITS, J. J., PEURIFOY, P. V., and WOODS, L. A.: Anal. Chem. 42, 351 (1970)
15. OKUMURA, T., and KADONO, T.: U.S. Patent 3, 829, 205 (1974).
16. Shionogi Co. Ltd.: British Patent 1, 390, 258 (1975).
17. RANNÝ, M.: TLC-FID, A New Technique in Quantitative Thin-Layer Chromatography (in Czech), Academia, Prague, 1984.
18. GIDDINGS, J. C.: Dynamics of Chromatography, Marcel Dekker, New York, 1965.
19. SNYDER, L. R.: Principles of Adsorption Chromatography. Marcel Dekker, New York, 1968.
20. NOVÁK, J., JANÁK, J., and WIČAR, S.: Basic Processes in Chromatography. Liquid Chromatography (Z. Dejl, K. Macek, and J. Janák, eds.). J. Chromatogr. Library, Vol. 3., Elsevier, Amsterdam, 1975.
21. GEISS, F.: Die Parameter der Dünnschichtchromatographie, F. Vieweg und Sohn, Braunschweig, 1972.
22. MARTIN, A. J. P., and SYNGE, R. L. M.: Biochem. J. 35, 1358 (1941).
23. WILKE, C. R., and CHANG, P.: AIChE J. 1, 264 (1955).
24. BELENKII, B. G., NESTEROV, V. V., GANKINA, E. S., and SMIRNOV, M. M.: J. Chromatogr. 31, 360 (1967).
25. VAN DEEMTER, J. J., ZUIDERWEG, F. J., and KLINKENBERG, A.: Chem. Eng. Sci. 5, 271 (1956).

26. KNOX, J. H.: J. Chromatogr. Sci. *15*, 353 (1977).
27. DELIGNY, C. L., and REMIJNSE, A. G.: J. Chromatogr. *33*, 242 (1968).
28. KAISER, R. E.: Chromatographia *9*, 463 (1976).
29. SCHMUTZ, H. R.: New Aspects in Quantitative HPTLC. In reference 7, p. 246.
30. TANAKA, M., ITOH, T., and KANEKO, H.: Yukagaku *25*, 263 (1976).
31. TANAKA, M., ITOH, T., and KANEKO, H.: Lipids *15*, 872 (1980).
32. BANERJEE, A. K., RATNAYAKE, W. M. N., and ACKMAN, R. G.: Lipids *20*, 121 (1985).
33. NEWMAN, J.: Chromarod Development Conditions and Scanning Parameters. NHA, 1981.*
34. HIRAYAMA, O., and MORITA, K.: Agric. Biol. Chem. *44*, 2217 (1980).
35. ACKMAN, R. G.: Methods Enzymol. 72, 205 (1981).
36. RANNÝ, M., SEDLÁČEK, J., BLÁHOVÁ, M., and POKORNÝ, J.: Unpublished results.
37. VAN AERDE, P., VAN SEVEREN, R., and BRAECKMAN, P.: Farm. Tijdschr. Belge *56*, 301 (1979).
38. Adjustment of Chromarod Activity by the Vacuum Drying Process. NHA, 1982.
39. TRAPPE, W.: Biochem. Z. *305*, 150 (1940); *306*, 316 (1940).
40. SNYDER, L. R.: Eight International Symposium on Gas Chromatography, Institute of Petroleum, Dublin, 1970.
41. HALPAAP, H., and RIPPHAHN, J.: HPTLC Development, Data and Results. In reference 7, p. 126.
42. Optimizing Reproducibility with the Iatroscan TLC-FID Analyzer. NHA, 1982.
43. Chromarod Conditioning and Cleaning. NHA, 1984.
44. STERNBERG, J. C., GALLAWAY, W. S., and JONES, D. T. L.: The Mechanism of Response of Flame Ionization Detector. Gas Chromatography, ISA Proceedings 1961. Brenner, N., Callen, J. E., and Weiss, M. D. (eds.). Academic Press, New York, 1962.
45. BOČEK, P., and JANÁK, J.: Chromatogr. Rev. *15*, 111 (1971).
46. ŠEVČÍK, J.: Detectors in Gas Chromatography. Elsevier, Amsterdam, 1976.
47. TANAKA, M., ITOH, T., KANEKO, H., KATO, O., ISHII, J., YU, R. K., OGAWA, T., and KAWANISHI, P.: A Test to Improve the Sensitivity of the Iatroscan. Oral Presentation at the 20th Annual Meeting of the Japan Oil Chemists' Society, 1981. English translation, NHA, 1981.
48. BRADLEY, D. M., RICKARDS, C. R., and THOMAS, N. S. T.: Clin. Chim. Acta *92*, 293 (1979).
49. RUSSEV, P., GOUGH, T. A., and WOOLLAM, C. J.: J. Chromatogr. *119*, 461 (1976).
50. KARAGÖZLER, A. E., and SIMPSON, C. F.: J. Chromatogr. *150*, 329 (1978).
51. ROBERTS, J. E., and HILL, H. H. Jr.: J. Chromatogr. *176*, 1 (1979).
52. RANNÝ, M., BLÁHOVÁ, M., TRUCHLÍK, Š., and ZBIROVSKÝ, M.: Unpublished results.
53. PATTERSON, P. L.: Lipids *20*, 503 (1985).
54. CRANE, R. T., GOHEEN, S. C., LARKIN, E. C. and RAO, G. A.: Lipids *18*, 74 (1983).
55. BANERJEE, A. K., RATNAYAKE, W. M. N., and ACKMAN, R. G.: J. Chromatogr. *319*, 215 (1985).
56. ZBIROVSKÝ, M., RANNÝ, M., BLÁHOVÁ, M., WITEK, S., and OSWIECIMSKA, M.: Proceedings of the Symposium on Chemistry of Pesticides. Lewin Klodzki, 1981. Wydawnictwo Politechniki Wroclawskiej, Wroclaw, 1981.
57. DELMAS, R. P., PARRISH, C. C., and ACKMAN, R. G.: Anal. Chem *56*, 1272 (1984).
58. ECKSCHLAGER, K.: Errors Measurement and Results in Chemical Analysis. Van Nostrand Reinhold, London, 1969.
59. TATARA, T., FUJII, T., KAWASE, T., and MINAGAWA, M.: Lipids *18*, 732 (1983).
60. Plotting Calibration Curves. Technical Instruction OpM/148. Iatron Labs. Inc., Tokyo, 1984.

---

* References NHA are Technical Instructions of the Newman-Howells Associates, Ltd., Winchester, Hants SO23 9NB, England.

61. Mareš, P., Ranný, M., Sedláček, J., and Skořepa, J.: J. Chromatogr., Biomed. Appl. *275*, 295 (1983).
62. Nomenclature of Lipids. IUPAC-IUB Commission on Biochemical Nomenclature, Hoope--Seyler's Z. Physiol. Chem. *38*, 517 (1977).
63. Hawthorne, J. N., and Ansell, G. B. (eds.): Phospholipids. Elsevier Biomedical Press, Amsterdam, 1982
64. Christie, W. W.: Lipid Analysis. Pergamon Press, Oxford, 1982.
65. Kanfer, J. N., and Hakamori, S.: Sphingolipids Biochemistry. Handbook of Lipid Research (Hanahan, J., ed.) Vol. 3, Plenum Press, New York, 1983.
66. Curtius, H. C., Roth, M. (eds.): Clinical Biochemistry. W. D. Gruyter, New York, 1974.
67. Henry, R. J., Cannon, D. C., and Winkelman, J. W.: Clinical Chemistry. Harper and Row, New York, 1974.
68. Kuksis, A., Marai, L., and Gornall, D. A.: J. Lipids Res. *8*, 352 (1967).
69. Kuksis, A., Myher, J. J., Marai, L., and Geher, K.: J. Chromatogr. Sci. *13*, 423 (1975).
70. Mareš, P., Tvrzická, E., and Tamchyna, V.: J. Chromatogr., Biomed. Appl. *146*, 241 (1978).
71. Mareš, P., Tvrzická, E., Skořepa, J., and Tamchyna, V.: J. Chromatogr., Biomed. Appl. *164*, 331 (1979).
72. Kawai, T., Hasunuma, S., Nakano, E., Sakurabayashi, I., Okkubo, N., Yoshioka, S., and Ishii, J.: Jap. J. Clin. Pathol. *19*, 293 (1971).
73. Nakano, E., Sakurabayashi, I., Hasunuma, S., Kawai, I., Tsuchia, T., and Okkuna, N.: Jap. J. Clin. Pathol. *20*, 186 (1972).
74. Tokunaga, M., Ando, S., and Ueda, N.: Proc. Jap. Conf. Biochem. Lipids *15*, 195 (1973).
75. Ueda, K., Itoh, K., Tejima, T., Kano, M., and Tadano, J.: Jap. J. Med. Technol. *19*, 639 (1975).
76. Inoue, T.: Physiol. and Ecol. *16*, No. 1 (1975).
77. Vandamme, D., Vankerckhoven, G., Vercaemst, R., Soetewey, F., Blaton, V., Peeters, H., and Rosseneu, M.: Clin. Chim. Acta *89*, 231 (1978).
78. Vandamme, D., Blaton, V., and Peeters, H.: J. Chromatogr. *145*, 151 (1978).
79. Herslöf, B.: Application of TLC-FID in Lipid Analysis. Advances in the Biochemistry and Physiology of Plant Lipids (Appelqvuist L. A., and Liljenberg, C., eds.) p. 301. Elsevier North--Holland Biomedical Press, Amsterdam, 1979.
80. Van Tornout, P., Vercaemst, R., Caster, H., Lievens, M. J., DeKeersgieter, W., Soetewey, F., and Rosseneu, M.: J. Chromatogr., Biomed. Appl. *164*, 222 (1979).
81. Itoh, T., Tanaka, M., and Kaneko, H.: Flame Ionization Detection System for TLC of Lipids. Thin Layer Chromatography (Touchstone S. C., and Rogers, D., eds.), p. 536, Wiley, New York 1981.
82. Tadano, J., Niwa, M., Ueda, H., and Kano, M.: Tokai J. Exp. Clin. Med. (Tokyo) *4*, 15 (1979).
83. Tadano, T.: Med. Technol. (Tokyo) *8*, 1203 (1980).
84. Zahler, P., and Niggli, V.: Methods in Membrane Biology, Vol. 8 (Korn E. D., ed.). Plenum Press, New York, 1970.
85. Folch, J., Ascoli, I., Lees, M., Meath, J., and Lebaron, F. N.: J. Biol. Chem. *191*, 833 (1951).
86. Rouser, G., Simon, G., and Kritchevsky, G.: Lipids *4*, 599 (1969).
87. Stearns, E. M. Jr., and Morton, W. T.: Lipids *8*, 668 (1973).
88. Rouser, G., Kritchevsky, G., and Yamamoto, A.: Lipid Chromatographic Analysis, Vol. 3 (Marinetti, G. V., ed.). Marcel Dekker, New York 1976.
89. Bjerve, K. S., Daae, L. N. W., and Bremer, J.: Anal. Biochem. *58*, 238 (1974).
90. Dawson, R. M. C.: Biochem. J. *75*, 45 (1960).
91. Roughan, P. G., Slack, C. R., and Holland, R.: Lipids *13*, 497 (1978).

184

92. NICHOLS, B. W.: Biochem. Biophys. Acta 70, 417 (1963).
93. HARA, A., and RADIN, N. S.: Anal. Biochem. 90, 420 (1978).
94. BLIGH, E. G., and DYER, W. J.: Can. J. Biochem. Physiol. 37, 911 (1959).
95. WELLS, M. A., and DITTMER, J. C.: Biochemistry 2, 1259 (1963).
96. WUTHIER, R. E.: J. Lipids Res. 7, 558 (1966).
97. LUCAS, C. C., and RIDOUT, J. H.: Prog. Chem. Fats 10, 1 (1970).
98. SHISHIDO, N., ISOBE, T., HORI, I., and UDAKA, K.: J. Tox. Sci. 1976, p. 93.
99. Analysis of Serum Lipids by the Iatroscan. Technical Instruction 132-OpM. Iatron Labs. Inc., Tokyo 1982.
100. DODGE, J. T., MITCHELL, C., and HANAHAN, D. J.: Arch. Biochem. Biophys. 100, 119 (1963).
101. TAKEMOTO, Y.: Kawasaki Med. J. 6, 1 (1980).
102. ROSE, H. G., and OKLANDER, M.: J. Lipid Res. 6, 428 (1965).
103. TSUCHIYA, Y., and SUGAI, H.: Biochem. Med. 28, 256 (1982).
104. NAGAYAMA, T., KUDO, M., NONAKA, H., AOYAMA, A., AKIMA, M., and FUKUNAGA, N.: Proceedings of the International Symposium on the Leucodystrophy and Allied Diseases, p. 27. Kyoto, 1981.
105. TOKUNO, K., MITANI, K., TANAKA, H., ITO, K., WATANABE, C., MURATA, K., OGINO, H., KINO, M., and MATSUMURA, T.: Rinsho Kensa 27, 322 (1983). English translation, Iatron Labs. Inc. No. 146/16, Tokyo 1983.
106. PONTHIEU, A. M., PORCHET, N., FRUCHART, J. C., SEZILLE, G., DEWAILLY, P., CODACCIONI, X., and DELECOUR, M.: Clin. Chem. 25, 31 (1979).
107. HIRAMATSU, K., and ARIMORI, S.: J. Chromatogr. 227, 423 (1982).
108. NHA no. 7 (3, 1, 1977).
109. SIPOS, J. C., and ACKMAN, R. G.: J. Chromatogr. Sci. 16, 443 (1978).
110. CHRISTIE, W. W., and HUNTER, M. L.: J. Chromatogr. 171, 517 (1979).
111. Analysis of Lipids by the Iatroscan. Technical Instruction 134-OpM. Iatron Labs. Inc., Tokyo, 1982.
112. KRAMER, J. K. G., FOUCHARD, R. C., and FARNWORTH, E. R.: J. Chromatogr. 198, 279 (1980).
113. PARRISH, C. C., and ACKMAN, R. G.: Lipids 18, 563 (1983).
114. MILLS, G. L., TAYLAUR, C. E., and MILLER, A. L.: Clin. Chim. Acta 93, 173 (1979).
115. INNIS, S. M., and CLANDININ, M. T.: J. Chromatogr. 205, 490 (1981).
116. NHA no. 53 (20, 3, 1979).
117. BLÁHOVÁ, M., RANNÝ, M., and SEDLÁČEK, J.: Acta Univ. Carol. Med. 32, No. 1–2 (1986).
118. Iatroscan Analysis and HP 3390 A Integration of Lipids. NHA, 1981.
119. HIRAMATSU, K., NOZAKI, H., and ARIMORI, S.: J. Chromatogr. 182, 301 (1980).
120. LEA, C. H., RHODES, D. N., and STOLL, R. D.: Biochem. J. 60, 353 (1955).
121. WAGNER, H., HÖRHAMMER, L., and WOLFF, P.: Biochem. Z. 334, 175 (1961).
122. SEDLÁČEK, J., RANNÝ, M., BLÁHOVÁ, M., and RŮŽIČKA, V.: Acta Univ. Carol. Med. 32, No. 1–2 (1986).
123. FURUYA, T., NAGUMO, T., ITOH, T., and KANEKO H.: Proc. Jap. Conf. Biochem. Lipids 20, 5 (1978).
124. BYRNE, H., LOUGHREY, M., and LETTERS, R.: Proceedings of the 19th International Congress of European Brewery Convention (EBC), p. 659, London, 1983.
125. TANAKA, M., TAKASE, K., ISHII, J., ITOH, T., and KANEKO, H.: J. Chromatogr. 284, 433 (1984); Itoh, T., Tanaka, M. and Kaneko, H.: Lipids 20, 553 (1985).
126. KAITARANTA, J. K., and NICOLAIDES, N.: J. Chromatogr. 205, 339 (1981).
127. KUBOTA, G., MORITA, N., FUJIVO, T., and MORI, L.: Iatroscan TH 10 Method for Amniotic Fluid Surfactant Assay. Lecture at the Conference of the Japan Society of Obsterics and Gynaecology, Tokyo, 1979. English translation, NHA, 1980.

128. PEETERS, H.: The Biological Significance of Plasma Phospholipids: Phosphatidylcholine, Biochemical and Clinical Aspects of Essential Phospholipids (Peeters H., ed.). Springer-Verlag, New York, 1976.

129. FREDRICKSON, D. S., GOLDSTEIN, J. I., and BROWN, M. S.: The Familar Hyperlipoproteinemias in the Metabolic Basis of Inherited Diseases (Stanbury, J. B., Wyngaarden, J. B., and Fredrickson, D. S., eds.). McGraw-Hill Book Co., New York, 1978.

130. KMENT, M., MUSIL, J., VÍŠEK, V., and RANNÝ, M.: Acta Univ. Carol. Med. submitted for publication.

131. KUŽELA, L., MUSIL, J., VÍŠEK, V., and RANNÝ, M.: Acta Univ. Carol. Med., submitted for publication.

132. BUCKIOVÁ, D.: Dissertation. Faculty of Medicine and Hygiene, Charles University, Prague, 1984.

133. EVANS, W. M.: Plasma Membrane Isolation. Laboratory Techniques in Molecular Biology and Biochemistry (Work, T. S., ed.). Elsevier, Amsterdam, 1979.

134. SHATTIL, S. J., and COOPER, R. A.: J. Lab. Clin. Med. 89, 341 (1977).

135. TAKEMOTO, Y.: Kawasaki Med. J. 7, 195 (1981).

136. YAMAMOTO, S., TAKEMOTO, Y., YAMASHITA, S., OHASHI, K., and HIRANO, Y.: Nippon Syôkakibyô Gakkai Zassi 76, 2424 (1979).

137. CLANDININ, M. T., FOOT, M., and ROBSON, L.: Comp. Biochem. Physiol. 76B, 335 (1983).

138. AMANUMA-MUTO, K., KANASEKI, T., IMANAKA, T., OHKUMA, S., and TAKANO, T.: Biochem. Internat. 7, 107 (1983).

139. BÖYUM, A.: Scand J. Clin. Invest. Suppl. 97, 21 (1968).

140. ROGIERS, V.: J. Chromatogr. 182, 27 (1980).

141. FOOT, M., and CLANDININ, M. T.: J. Chromatogr. 241, 428 (1982).

142. HORROCKS, S. L. A.: J. Lipid Res. 9, 469 (1968).

143. YOSHIZUKA, N., OKAMOTO, K., and TAKASE, Y.: Kosho Kai Shi 5, 33 (1982).

144. ACKMAN, R. G., NASH, D., and McLACHLAN, J.: Proc. N. S. Inst. Sci. 29, 501 (1979).

145. HAZEL, J. R.: Lipids 20, 516 (1985).

146. KRAMER, J. K. G., FARNWORTH, E. R., and THOMPSON, B. K.: Lipids 20, 536 (1985).

147. KATSU, M., OKUTSU, S., TSUTSUMI, N., HASEGAWA, K., and INOUE, T.: J. Kagawa Nutrition Coll. 10, 9 (1979).

148. OKUTSU, S., KIMURA, K., TERUNUMA, R., TSUTSUMI, N., and HASEGAWA, K.: J. Kagawa Nutrition Coll. 10, 16 (1979).

149. FRUCHART, J. C., PONTHIEU, A., PORCHET, N., DEWAILLY, P., and SEZILLE, G.: Ann. Biol. Clin. 36, 149 (1978).

150. HALLMAN, M., KULOVICH, M., KIRKPATRICK, E., SUGARMAN, R. G., and GLUCK, L.: Amer. J. Obstet. Gynaecol. 125, 613 (1976).

151. TSAI, M. Y., and MARSHALL, S.: Clin. Chem. 25, 628 (1979).

152. GLUCK, L., KULOVICH, M., BORER, R. C., BRENNER, P. H., ANDERSON, G. G., and SPELLACY, W. N.: Amer. J. Obstet. Gynaecol. 109, 441 (1971).

153. FREER, D. B., and STATLAND, B. E.: Clin. Chem. 27, 1629 (1981).

154. SIMON, R. G.: Clin. Chem. 18, 315 (1972).

155. BADHAM, L. P., and WORTH, H. G. J.: Clin. Chem. 21, 1441 (1975).

156. HARVEY, H. R., RIGLER, M. W., and PATTON, J. S.: Lipids 20, 542 (1985).

157. KNAPP, R. D., and SHERRILL, B. C.: Lipoprotein Lipid Quantitation by Iatroscan. Lecture at the 75th Annual Meeting of the AOCS, Dallas, 1984. J. Amer. Oil Chem. Soc. 61, No. 4 (1984).

158. RAMIREZ, F., MARECEK, J. F., and YEMUL, S. S.: J. Org. Chem. 48, 1417 (1983).

159. DROZDZ, M., KUCHARZ, E., and SZYJA, J.: Environ. Res. 13, 369 (1977).

160. KOBAYASHI, T., and KUBOTA, K.: Chemosphere 9, 777 (1980).

161. LERCKER, G., FREGA, N., CONTE, L. S., and CAPPELA, P.: Riv. Ital. Sostanze Grasse 58, 324 (1981).

186

162. Domnas, A. J., Warner, S. A., and Johnson, S. L.: Lipids *18*, 87 (1983).
163. Uji, A.: J. Pharm. Soc. Japan *95*, 214 (1975).
164. Masui, T., and Fujishima, M.: Kanzo *20*, 976 (1979).
165. Masui, T., and Fujishima, M.: Kanzo *20*, 1299 (1979).
166. Masui, T., and Fujishima, M.: Kanzo *21*, 233 (1980).
167. Beke, R., De Weerdt, G. A., and Barbier, F.: J. Chromatogr. *193*, 504 (1980).
168. Eneroth, P., and Sjövall, J.: Methods Enzymol. *15*, 237 (1969).
169. Dupont, J., Oh, S. Y., and Janssen, P.: Tissue Distribution of Bile Acids. Bile Acids (Nair, P. P., and Kritchevsky, D., eds.), Vol. 3. Plenum Press, New York 1976.
170. Street, J. M., Trafford, D. J. H., and Makin, H. L. J.: J. Lipid Res. *24*, 491 (1983).
171. Mingrone, G., Greco, A. V., Boniforti, L., and Passi, S.: Lipids *18*, 90 (1983).
172. Eneroth, P., and Sjövall, J.: Extraction, Purification and Chromatographic Analysis of Bile Acids in Biological Materials. Bile Acids (Nair, P. P., and Kritchevsky, D. eds.), Vol. 1. Plenum Press, New York 1971.
173. NHA no. 3 (8, 9, 1976).
174. Rigler, M. W., Leffert, R. L., and Patton, J. S.: J. Chromatogr. *277*, 321 (1983).
175. Set of TLC-FID Chromatograms. NHA, October 1980.
176. Okumura, T., Kadano, T., and Iso'o, A.: J. Chromatogr. *108*, 329 (1975).
177. Urinary Steroids NHA, 1980.
178. Waldi, D.: Klin. Wochschr. *40*, 827 (1962).
179. Namba, T., Yoshioka, S., and Newman, J.: Laboratory Equipment Digest, January 1980.
180. NHA no. 44 (29, 11, 1977).
181. Instrument Applications No. 2. Iatron Labs. Inc., Tokyo 1977.
182. NHA no. 19 (26, 7, 1977).
183. Jacini, G., and Fedeli, E.: Fette, Seifen, Anstrichmittel *77*, 1 (1975).
184. Ranný, M., Bláhová, M., and Novotný, L.: Unpublished results.
185. NHA no. 9 (10, 2, 1977).
186. NHA no. 66 (15, 8, 1978).
187. Hušek, P., and Macek, K.: J. Chromatogr. *113*, 139 (1975).
188. Set of TLC-FID Chromatograms. NHA, April 1980.
189. TLC-FID of Arnica, Foin, Eichenmoss and Tolu Resins. NHA, 1980.
190. Novotný, L., Herout, V., and Šorm, F.: Coll Czech. Chem. Commun. *29*, 2182 (1964).
191. Novotný, L., Samek, Z., Harmatha, J., and Šorm, F.: Coll. Czech. Chem. Commun. *34*, 336 (1969).
192. Yamagushi, T., Kaneshima, H., and Yamagishi, T.: Eisei Kagaku *21*, 286 (1975).
193. Ono, M., Shimamine, M., and Takahashi, K.: Eisei Shikensho Hokoku *96*, 67 (1978).
194. Saito, T., Kaneshima, H., and Okada, M.: Report of Hokkaido Institute of Public Health *33*, 131 (1983).
195. Takase, Y., and Yoshioka, S.: Separation of Sulphanilic Acid and Sulphonamides. Iatron Labs. Inc., Tokyo, 1977.
196. An Analytical Procedure for Handling Small Amounts of Substances by TLC-FID. Iatron Labs. Inc., Tokyo, 1982.
197. Pohle, W. D., and Mehlenbacher, V. C.: J. Amer. Oil Chem. Soc. *27*, 54 (1950).
198. Dutton, J. H.: J. Amer. Oil Chem. Soc. *32*, 652 (1955).
199. Aylward, F., and Wood, P. D. S.: Nature *177*, 146 (1956).
200. Dieffenbacher, A., and Bracco, U.: J. Amer. Oil Chem. Soc. *55*, 642 (1978).
201. Quinlin, P., and Weiser, H. J. Jr.: J. Amer. Oil Chem. Soc. *35*, 325 (1958).
202. Watts, R., and Dils, R.: J. Lipid Res. *10*, 33 (1969).
203. Blum, J., and Koehler, W. R.: Lipids *5*, 601 (1970).
204. Aitzetmüller, K.: J. Chromatogr. *113*, 231 (1975).

205. HAMMOND, E. W.: J. Chromatogr. *203*, 397 (1981).
206. RANNÝ, M., SEDLÁČEK, J., MAREŠ, E., SVOBODA Z., and SEIFERT, R.: Seifen, Oele, Fette, Wachse *109*, 219 (1983).
207. GANTOIS, E., MORDRET, F., LEBARBACHON, N., and BARBATTI, C.: Rev. Franc. Corps Gras *24*, 467 (1977).
208. NHA no. 46 (5, 1, 1979).
209. PORÉ, J., and RASORI, I.: Parfums, Cosmet. Arômes *46*, 31 (1982).
210. BINDLER, F., LAUGEL, P., and HASSELMAN, M.: Deutsche Lebensmittel-Rundschau *75*, 111 (1979).
211. PORÉ, J., HOUIS, J. P., and RASORI, I.: Rev. Fr. Corps Gras *28*, 111 (1981).
212. SEDLÁČEK, J.: Dissertation. Institute of Chemical Technology, Prague 1986.
213. LJUSBERG-WAHREN, H., HERSLÖF, M., and LARSSON, K.: Chem-Phys. Lipids *33*, 211 (1983).
214. GUHA, O. K., and JANÁK, J.: J. Chromatogr. *68*, 325 (1972).
215. TANAKA, M., ITOH, T., and KANEKO, H.: Yukagaku *28*, 96 (1979).
216. ACKMAN, R. G., and SEBEDIO, J. L.: Biological Filtration of Partially Hydrogenated Fish Oil in Nonhuman Primate Species. Paper presented at the AOCS Meeting, New Orleans, 1981.
217. SEBEDIO, J. L., FARQUHARSON, T. E., and ACKMAN, R. G.: Lipids *17*, 469 (1982).
218. SEBEDIO, J. L., FARQUHARSON, T. E., and ACKMAN, R. G.: Lipids *20*, 555 (1985).
219. SEBEDIO, J. L., and ACKMAN, R. G.: J. Chromatogr. Sci. *19*, 552 (1981).
220. TANAKA, M., ITOH, T., and KANEKO, H.: A New Method of Lipid Analysis in View of Its Unsaturation with the Iatroscan TH-10. Paper presented at the 15th Meeting of the Japan Oil Chem. Soc., October 1976.
221. Analysis of Triglyceride Molecular Species by Silver Nitrate Impregnated Chromarods. Technical Instruction, Iatron Labs, Inc., Tokyo, 1982.
222. Method for the Preparation of Silver Nitrate Impregnated Chromarods. Technical Instruction, Iatron Labs, Inc., Tokyo, 1983.
223. NHA no. 14 (27, 6, 1977).
224. SEBEDIO, J. L., and ACKMAN, R. G.: J. Amer. Oil Chem. Soc. *60*, 1992 (1983).
225. PETERSSON, B.: J. Chromatogr. *242*, 313 (1982).
226. ISOBE, T., SEINO, H., and WATANABE, S.: Study of the Lipase Hydrolysis of Castor Oil. NHA, 1979.
227. KANEKO, H., HOSOHARA, M., TANAKA, M., and ITOH, T.: Lipids *11*, 837 (1976).
228. REDGRAVE, T. G., and JEFFERY, F.: Lipids *16*, 626 (1981).
229. SIBBALD, I. R., and KRAMER, J. K. G.: Poultry Sci. *59*, 1505 (1980).
230. RAO, G. S. L., WILLISON, J. H. M., and RATNAYAKE, W. M. N.: Plant Physiol. *75*, 716 (1984).
231. TOYOMIZU, M., HANAOKA, K., and NAKAMURA, T.: Bull. Jap. Soc. Sci. Fisheries *46*, 1011 (1980).
232. LINKO, R., and KAITARANTA, J. K.: J. Sci. Agric. Soc., Finland *52*, 423 (1980).
233. KAITARANTA, J. K., and ACKMAN, R. G.: Comp. Biochem. Physiol. *69B*, 725 (1981).
234. ACKMAN, R. G.: Chem. Ind., October 1981, p. 715.
235. KAITARANTA, J. K.: J. Food Technol. *17*, 87 (1982).
236. PATTON, J. S., and BURRIS, J. E.: Marine Biol. *75*, 131 (1983).
237. KELLOGG, R. B., and PATTON, J. S.: Marine Biol. *75*, 137 (1983).
238. VETTER, R. D., and HODSON, R. E.: Can J. Fish Aquat. Sci. *40*, 627 (1983).
239. PATTON, J. S., BATTEY, J. F., RIGLER, M. W., PORTER, J. W., BLACK, C. C., and BURRIS, J. E.: Marine Biol. *75*, 121 (1983).
240. KAITARANTA, J. K.: Fish Roe Lipids and Lipid Hydrolysis in the Processed Roe of Certain Salmonidae Fish as Studied by Novel Chromatographic Techniques. Research Report 14/1981, Technical Research Centre of Finland. English translation, NHA, 1983.
241. ELDRIDGE, M. B., JOSEPH, J. D., TABERSKI, K. M., and SEABORN, G. T.: Lipids *18*, 510 (1983).

188

242. PARRISH, C. C., and ACKMAN, R. G.: J. Chromatogr. *262*, 103 (1983).
243. PARRISH, C. C., and ACKMAN, R. G.: Lipids *20*, 521 (1985).
244. KAITARANTA, J. K.: J. Sci. Food Agric. *31*, 1303 (1980).
245. RATNAYAKE, W. N., and ACKMAN, R. G.: Lipids *14*, 795 (1979).
246. KAITARANTA, J. K., and KE, P. J.: J. Amer. Oil. Chem. Soc. *58*, 710 (1981).
247. ZEMAN, I., RANNÝ, M., and WINTEROVÁ, L.: J. Chromatogr., 354, 282 (1986).
248. HARVEY, H. R., and PATTON, J. S.: Anal Biochem. *116*, 312 (1981).
249. HIRAYAMA, O., and MORITA, K.: Agric. Biol. Chem. *44*, 2217 (1980).
250. DEGUCHI, K., KAWASHIMA, S., ICHIO H., and UETA, N.: J. Biochem. *85*, 1519 (1979).
251. RANNÝ, M., ŠILHÁNEK, J., BRADÍKOVÁ, A., SEIFERT, R., and ZBIROVSKÝ, M.,: Tenside Detergents *13*, 77 (1976).
252. RANNÝ, M., and LIST, J.: Acta Univ. Carol. Med. 32, No. 1–2 (1986).
253. RANNÝ, M., ŠILHÁNEK, J., BRADÍKOVÁ, A., SEIFERT, R., and ZBIROVSKÝ, M.: Tenside Detergents *14*, 246 (1977).
254. ACKMAN, R. G., and WOYEWODA, A. D.: J. Chromatogr. Sci. *17*, 514 (1979).
255. DUPLESSIS, L. M., and PRETORIUS, H. E.: J. Amer. Oil Chem. Soc. *60*, 1261 (1983).
256. DAUN, J. K., and DECLERCQ, D. R.: Use of Thin Layer Chromatography with Flame Ionization Detection to Determine Acetone Insolubles in Crude and Degummed Rapeseed Oils. Proceedings of the Symposium on the Analytical Chemistry of Rapeseed Oil and Its Products (Daun, J. K., McGregor, D. I., and McGregor, E. E., eds.). The Canola Council of Canada, Winnipeg, 1980
257. NHA no. 51 (6, 3, 1979).
258. ČAPEK, K., VYDRA, T., ČAPKOVÁ, J., RANNÝ, M., BLÁHOVÁ, M., and SEDMERA, P.: Coll. Czech. Chem. Commun., 50, 1039 (1985).
259. KONDO, Y.: Agric. Biol. Chem. *41*, 2481 (1977).
260. KONDO, Y.: Carbohydrate Res. *103*, 154 (1982).
261. RANNÝ, M., and ČAPEK, K.: Unpublished results.
262. ČAPEK, K., and JARÝ, J.: Coll. Czech. Chem. Commun. *38*, 2518 (1973).
263. KANESHIMA, H., OGAWA, H., YAMAGISHI, T., KINOSHITA, Y., and MORI, M.: Hokkaidoritsu Eisei Kenkyusho Ho *24*, 64 (1974).
264. NHA (26, 1, 1981).
265. NHA no. 50 (2, 3, 1979).
266. FUJII, T., TANAKA, R., SASAI, K., and TANAKA, T.: Microquantitative Analysis of Surface Agents Using Iatroscan TH-10. Paper presented at the 14th Annual Meeting of the Japan Oil Chemists's Society, Tokyo, 1975. NHA, 1977.
267. NHA no. 59 (18, 4, 1979).
268. COZZOLI, O.: Riv. Ital. Sostanze Grasse *57*, 136 (1980).
269. NHA no. 20 (29, 7, 1977).
270. NHA no. 62 (26, 7, 1979).
271. Iatroscan TH-10 Analyser Mark III Manual, Iatron Labs. Inc., Tokyo, 1979.
272. NHA no. 33 (18, 4, 1978).
273. ESCOTT, R. E. A., BRINKWORTH, S. J., and STEEDMAN, T. A.: J. Chromatogr. *282*, 655 (1983).
274. ANDO, Y., and HAYASHI, M.: Yukagaku *31*, 372 (1982).
275. SCOTT, H.; Removal of Organic Soil from Fibrous Substrates. In Detergency (Cutler, W. G., and Davies, R. C., eds.) Marcel Dekker, New York, 1972.
276. ANDREE, H., MÜLLER, C. W. R., and SCHMID, R. D.: J. Appl. Biochem. *2*, 218 (1980).
277. NHA no. (76, 21, 1, 1980).
278. YOKONO, T., KANAKIYO, J., OZAWA, H., and SANADA, Y.: J. Petrol. Inst. *26*, 1 (1983). English translation, Iatron Labs. Inc., Tokyo, 1983.
279. YOKOYAMA, S., UMEMATSU, J., INOUE, K., KATOH, T., and SANADA, Y.: Chemical Structure

of Coal Hydrogenation Liquids; Compounds Type Estimation by TLC-FID. Proceedings Internationale Kohlenwissenschaftliche Tagung, Düsseldorf, 1981. Verlag Glückauf, Essen, 1981. (Fuel 63, 984 (1984))

280. NHA no. 12 (26, 4, 1977); NHA no. 29 (7, 2, 1978).

281. POIRIER, M. A., RAHIMI, P., and AHMED, S. M.: J. Chromatogr. Sci. 22, 116 (1984).

282. KESSLER, H., and MÜLLER, E.: J. Chromatogr. 24, 469 (1966).

283. SUZUKI, Y., and TAKEUCHI, T.: Studies on Quantitative TLC by Means of a Hydrogen FID; II. Rapid Analysis of Fuel Oil Constituents. Paper presented at the 21st Annual Meeting of the Japan Society for Analytical Chemistry. English translation, NHA, 1977.

284. RAY, J. E., OLIVER, K. M., and WAINWRIGHT, J. C.: The Application of the Iatroscan TLC Technique to the Analysis of Fossil Fuels. Petroanalysis 81, IP Symposium, London, 27—29 October, 1981. Heyden and Son, London, 1981.

285. SELUCKY, M. L.: Anal. Chem. 55, 141 (1983); Lipids 20, 546 (1985).

286. FARCASIU, M.: Fuel 56, 9 (1977).

287. BODEN, H., and ROUSSEL, R.: A Method for Determining Polycyclic Aromatic Hydrocarbon for Coal Tar Pitch Volatiles. Internat. Environ. Safety News, June 1973, p. 7.

288. SUATONI, J. C., and SWAB, R. E.: J. Chromatogr. Sci. 13, 361 (1975).

289. POIRIER, M. A., and GEORGE, A. E.: J. Chromatogr. Sci. 21, 331 1983.

290. POIRIER, M. A., and GEORGE, A. E.: Energy Sources 7, 151 (1983).

291. Coal Tar Pitch Volatiles. Instrument Appl. No. 136, Iatron Labs., Inc., Tokyo, 1982.

292. UEDA, K., ITOH, K., TEJIMA, T., KANO, M., and TADANO, J.: Jap. J. Med. Technol. 19, 639 (1975).

293. SAWATSKI, H., GEORGE, A. E., SMILEY, G. T., and MONTGOMERY, D. S.: Fuel 55, 16 (1976).

294. OBUCHI, A., AOYAMA, H., OHI, A., and OHUCHI, H.: J. Chromatogr. 288, 187 (1984).

295. HOUIS, J. P., RASORI, I., and PORÉ, J.: TLC for Qualitative and Quantitative Analysis of Fats in Leathers. NHA, 1980.

296. HOUIS, J. P., ASSIMACOPOULOS, E., RASORI, I., and PORÉ, J.: The Application of TLC-FID to the Control of Industrial Emulsions. NHA, 1980.

297. PORÉ, J., and NOEL, J. P.: Rev. Tech. Ind. Cuir 72, 65 (1980).

298. PORÉ, J., and RASORI, I.: Noveaux developpements de la C. C. M./D. I. F. dans l'analyse des corps gras et leurs derivés. NHA, 1983.

299. PORÉ, J.: J. Soc. Leather Technol. Chem. 66, 41 (1982).

300. HOUIS, J. P., ASSIMACOPOULOS, E., RASORI, I., and PORÉ, J.: Rev. Fr. Corps Gras 27, 61 (1980).

301. NEWMAN, J. M.: Quality Control of Industrial Emulsions and Lubricants. NHA, 1981.

302. BELENKII, B. G., and GANKINA, E. S.: J. Chromatogr. 141, 13 (1977).

303. INAGAKI, H.: Adv. Polymer Sci. 24, 189 (1977).

304. GANKINA, E. S., and BELENKII, B. G.: Metody Anal. Kontrolya Kach. Prod. Khim. Prom-sti 1978, p. 21.

305. MIN, T. I., MIYAMOTO, T., and INAGAKI, H.: Rubber Chem. Tech. 50, 63 (1977).

306. MIN, T. I.: Pollimo 2, 146 (1978) (Korean).

307. CTPB and HTPB Resins and Analysis of Acrylic Resin Coatings. NHA, 1980.

308. HADAG, K. D.: New Developments in Chromatographic Characterization and Quality of Composite Materials. NHA, 1979.

309. MIN, T. I.: Newer Applicabilities of TLC for Polymer Characterization. Thesis (in English), Kyoto University, 1977.

310. MIN, T. I., KLEIN, A., EL-AASSER, M. S., and VANDERHOFF, J. W.: Organic Coatings and Applied Polymer Science Proceedings, 46, 314 (1982). ACS 183rd National Meeting, Las Vegas, Nevada, March–April 1982. (J. Polymer Sci 21, 2845 (1983)).

311. MIN, T. I., and INAGAKI, H.: Polymer 21, 309 (1980).

190

312. NHA no. 47 (2, 2, 1979).
313. NHA (3, 2, 1978).
314. RANNÝ, M., ZBIROVSKÝ, M., BLÁHOVÁ, M., RŮŽIČKA, V., and TRUCHLÍK, Š.: J. Chromatogr. *247*, 327 (1982).
315. MISSALA, I., and CZULIŇSKA, D.: Chem. analit. *13*, 23 (1968).
316. GRUCA, M., JANKO, Z., and KOTARSKI, A.: Chem. analit. *16*, 91 (1971).
317. ĎULÁK, K., KOVÁČ, J., and RAPOŠ, P.: J. Chromatogr. *31*, 354 (1967).
318. VÝTOH, P., MICHÁLEK, M., ŠUSTEK, J., and BATORA, V.: J. Pesticide Sci. *5*, 171 (1974).
319. SAHA, P. K., HATAKEDA, K., KATO, T., YAMANAKA, S., and TAKAHASHI, N.: Japan J. Crop. Sci. *50*, 382 (1981).
320. Read, H.: Lipids *20*, 510 (1985).
321. A Specific Detector for Nitrogen and Halogen Compounds in Thin Layer Chromatography on Coated Quartz Rods. NHA, 1984.
322. In the U.S.A. by Ancal Inc., 1530 Bayview Heights Drive Los Osos, CA 93402; outside the USA, by Newman-Howells Associates Ltd., Wolvesey Palace, Winchester, Hants SO23 9NB, England.
323. Newman, J. M.: Lipids *20*, 501 (1985).
324. Newman J. M.: Private communication (1985).
325. Kaimal, T. N. B., and Shantha, N. C.: J. Chromatogr. 288, 177 (1984).
326. Kramer, J. K. G., Fouchard, R. C., and Farnworth, E. R.: J. Chromatogr. 351, 571 (1986).
327. Kramer, J. K. G., Thompson, B. K., and Farnworth, E. R.: J. Chromatogr. 355, 221 (1986).
328. Iatroscan TH-10 Analyser Mark IV Manual. Iatron Labs. Inc., Tokyo, 1986.

Index